Traditional Livestock Production

Carlos Palacios Riocerezo
Manuel Gonzalez Ronquillo
Editors

Traditional Livestock Production

Adaptation to Organic Production Systems

Springer

Editors
Carlos Palacios Riocerezo
Area of Animal Production, Faculty of Agricultural and Environmental Sciences
University of Salamanca
Salamanca, Spain

Manuel Gonzalez Ronquillo
Facultad de Medicina Veterinaria y Zootecnia
Universidad Autonoma del Estado de Mexico
Toluca, Mexico

ISBN 978-3-031-88277-7 ISBN 978-3-031-88275-3 (eBook)
https://doi.org/10.1007/978-3-031-88275-3

This Springer imprint is published by the registered company Springer Nature Switzerland AG
The registered company address is: Gewerbestrasse 11, 6330 Cham, Switzerland

Preface

In recent years there has been an increase in the consumption of organic food in the world led by the United States, Germany and France, being a necessity to seek food free of hormones, pesticides, antibiotics and synthetic growth promoters. Likewise, consumers are paying more attention to the treatment of farm animals, that they are free grazing, that the food they receive does not compete with human food (e.g. cereal grains). The demand of these consumers concerned about the origin of the food they buy, promoting more sustainable systems, with healthier food without added substances in their production, respectful of the environment and with the certainty that the animals raised do so with the greatest guarantees of animal welfare, pushes producers to become certified in organic production. However, the traditional productions of small farmers, backyard economies, generally meet almost all the peculiarities necessary to be certified in organic production and that is why this editorial proposal combines the two production systems that help small farmers to continue with the traditional production system, dignifying their work and obtaining better economic returns.

This book provides a comprehensive overview of various aspects of sustainable agriculture, such as organic production systems focused on dairy farming of both cattle and sheep, indigenous poultry production, the use of insects in animal feed as a sustainable alternative, the production of lambs in organic grazing systems and the carbon footprint of organic livestock. It also addresses regulatory frameworks, challenges, opportunities and the importance of sustainability in agriculture, understanding why the animal products we consume should come from organic and not industrial sources.

This book is focused with an informative language to the future readers of this book, who can be farmers of traditional production systems, technicians who work with them and who facilitate improvements in their yields favouring better selling prices of their products, researchers in the sector and anyone interested in this type of emerging productions. Those who are interested in understanding how organic livestock farming works, what are the benefits of producing organic animal food, whether milk and its derivatives, as well as poultry meat and eggs, and feeding strategies that allow to be more sustainable with the environment, from extensive grazing and production of organic lamb meat, to the use of new alternative foods (i.e. insects) in organic food, understanding the legislation and regulations for its use.

This book, as a differentiator from other publications, brings together the situation of such disparate parts of the world as Patagonia (South America), Indonesia (Asia) and Europe from the point of view of producers and consumers. It also incorporates an open point of view, incorporating peasant and traditional productions as productions with great possibilities of certification and therefore producers of high-quality food.

Salamanca, Spain Carlos Palacios Riocerezo
Toluca, Mexico Manuel Gonzalez Ronquillo

Contents

1 **The Organic Production System: Regulation and Certification** 1
Évelyne Alcázar Marín

2 **European Organic Food Markets and Consumer Preferences** 19
Katrin Zander

3 **Carbon Footprint in Organic Livestock Farming** 29
Jaime Nieto de la Losa, Javier Plaza Martín, and Carlos Palacios Riocerezo

4 **Indigenous Poultry in Indonesia** 39
Wisje Lusia Toar, Laurentius J. M. Rumokoy, Carlos Palacios Riocerezo, and Manuel Gonzalez Ronquillo

5 **Insect Use in Organic Animal Production Systems** 55
Jaime Nieto de la Losa, Javier Plaza Martín, and Carlos Palacios Riocerezo

6 **Comparison of Milk Yield and Milk Composition Between Organic and Conventional Systems: A Meta-analysis and Meta-regression** 73
Juan Carlos Angeles-Hernandez, Rodolfo Vieyra Alberto, Ana Karen Zaldivar-Ortega, Navid Ghavipanje, Alfredo J. Escribano, and Manuel Gonzalez Ronquillo

7 **Organic Dairy Production in Europe. The Challenge of Conversion from Conventional Systems** 93
Ruth Rodríguez-Bermúdez, Marta Miranda, Marta López-Alonso, Ramiro Fouz, and Inmaculada Orjales

8 **Lamb Production in Patagonia** 127
Sergio Radic-Schilling, Francisco Sales, Raúl Lira, René Muñoz-Arriagada, Paulo Corti, Iván Ordoñez, Nilo Covacevich, Ricardo Echeverría, Camila Sandoval, and Jorge Ivelic-Sáez

9 **Sheep Production in Southern Chile Based on Pastures** 149
Francisco Canto, José Alfonso Abecia, María Eugenia Martínez, Rodrigo Morales, and Germán Holmberg

10 **Traditional Production System of Iberian Pig and Its Potential for Adaptation to Organic Livestock Farming** 167
Vicente Rodríguez-Estévez, Carolina Reyes-Palomo, Santos Sanz-Fernández, Pablo Rodríguez-Hernández, Ildefonso Caballero-Luna, and Cipriano Díaz-Gaona

Index 187

Chapter 1
The Organic Production System: Regulation and Certification

Évelyne Alcázar Marín

Abstract Organic production systems accounted for 2,1% of all agricultural systems in 2023 (FiBL & IFOAM – Organics International (2025): The World of Organic Agriculture. Frick and Bonn) and already cover more than 98.9 million hectares. It is a production system that has grown steadily since its start, it is regulated in 188 countries, including the European Union, and it is carried out by more than 4.3 million farmers worldwide.

Part of its success is based on having detected the problems of the conventional production system, which originated from the misnamed "Green Revolution". The "Green Revolution", based on the use of chemically synthesised fertilisers and pesticides, initially ignored all the problems derived from the production means alteration (soil, water and environment) and focused on productivity, which was very important in those times of post-war and food shortages. Organic production focused on the soil as the cornerstone of its functioning, as a response to avoiding contamination, to fight against desertification and against the loss of fertility and biodiversity, which were the most evident problems of conventional agriculture. For these reasons, the existing regulations for organic production focus on soil-based agriculture and livestock farming.

Back to the present, we now find that new agri-food policies are calling for sustainable food systems in the long term:

- Soil fertility problems are already affecting more than 60% of European agricultural soils and reducing the use of chemically synthesized fertilisers, recycling nutrients and restoring soil health is at the heart of the "farm to fork" strategy.
- Pesticide pollution problems affect the health of ecosystems (mainly by causing the disappearance of pollinators, soil and water contamination), but also the health of animals and humans. The European Commission is calling for a 50% reduction in the use of plant protection products and antibiotics in agriculture and livestock farming.

É. Alcázar Marín (✉)
Asociación ECOVALIA, Avda. Diego Martínez Barrio, nº10, 1º, Mod12, Seville, Spain
e-mail: ecovalia.internacional@ecovalia.org

C. Palacios Riocerezo, M. Gonzalez Ronquillo (eds.), *Traditional Livestock Production*, https://doi.org/10.1007/978-3-031-88275-3_1

- The major challenge of climate change demands that we rethink our agri-food production systems in terms of alternatives that are better adapted to extreme events, that reduce emissions and that guarantee food production in the long term. This climate emergency is a horizontal concern for European policies in all types of activity (transport, industry, energy, construction, etc.), not only in the agri-food sector, through the so-called "Green Deal".

Organic production is a response to these three problems and European policies set the objective that 25% of the land used agriculturally in Europe should be under this system of production as a way of making a transition in our agri-food systems. To achieve this, it is also necessary for a demand of this type of production and to identify it with a labelling on its products. The way to guarantee that these labelled products are complying with the requirements described in the different regulations is through certification systems. These certification systems ensure knowledge of the production standards, identify the aspects that must be controlled and define the supervision mechanisms that guarantee that the system works and is properly applied.

In this chapter we will learn about the different regulations and their implications for national, European and international marketing, and we will also go into details on what certification systems are and how they are enforced.

Keywords Organic farming · Organic European Regulation · Organic market · Sustainability · Soil Fertility · Biodiversity

1.1 The Context

1.1.1 Organic Farming in the World

The Organic production system is getting more a more relevance in the world, there is no one continent which does not count some surface in organic farming (3% in lowest level for Africa continent and 55% for the upper level in Oceania) (FiBL & IFOAM – Organics International 2024). This shows that Organic agriculture is a truly international movement. From vast fields in Europe and North America to smaller plots in Asia and South America, organic farming practices are taking root across the globe. This widespread adoption highlights the adaptability of organic principles to diverse climates, resources, and agricultural traditions.

While organic farming is present on every continent, the level of adoption varies considerably. Factors such as government support, consumer awareness, and economic development all play a role in shaping the organic market in different regions. Oceania, with its strong environmental consciousness and relatively high disposable income, boasts the highest percentage of organic farmland. Conversely, Africa faces challenges in infrastructure, access to knowledge, and market access, which contribute to its lower organic adoption rate (FiBL & IFOAM – Organics International 2024).

Despite those differences, since the start of global data collection over 25 years ago, global organic farmland has shown no signs of decline, with a consistent upward trend (FiBL & IFOAM – Organics International 2024). The steady increase in organic farmland reflects a growing appreciation for the benefits of organic practices. Consumers are increasingly seeking out organic products, driven by concerns about food safety, environmental sustainability, and animal welfare. This growth is mirrored in the organic market, which is also experiencing a significant and sustained rise. This rising demand incentivizes farmers to transition to organic methods, further expanding the global organic landscape.

The future of organic farming is promising. As environmental concerns mount and consumer demand for sustainable products continues to rise, we can expect organic agriculture to play an increasingly important role in global food systems. With continued research, innovation, and support for producers, organic farming has the potential to provide safe, healthy food for a growing population while protecting our planet's precious resources.

North America and Europe, the mainstream consumers regions in the world, had been developed relevant policies focusing on the sustainability problems in one hand and in the other on healthy diets, along to this chapter those topics are going to be detailed. The United States Department of Agriculture (USDA) has various initiatives promoting healthy eating patterns, such as the MyPlate guidelines and the National Organic Program (NOP) that sets standards for organic certification USDA MyPlate (https://www.choosemyplate.gov) and USDA National Organic Program (https://www.ams.usda.gov/about-ams/programs-offices/national-organic-program). The Canadian government offers similar programs like Canada's Food Guide and the Canadian Organic Regime: Health Canada – Canada's Food Guide, (https://food-guide.canada.ca/en/), Canadian Food Inspection Agency – Canadian Organic Regime (https://inspection.canada.ca/food-labels/organic-products/operating-manual/eng/1389199079075/1554143470958?chap=2). The European Union (EU) has developed a comprehensive strategy for sustainable food systems, including the Farm to Fork Strategy and the EU Organic Regulation: European Commission – Farm to Fork Strategy, (https://food.ec.europa.eu/horizontal-topics/farm-fork-strategy_en) European Commission – Biodiversity Strategy, (https://environment.ec.europa.eu/strategy/biodiversity-strategy-2030_en). These policies encourage sustainable agricultural practices, support organic farming conversion, and promote consumer awareness of healthy and environmentally friendly food choices. Those strategies had been recently replaced in their approach by the Vision for Agriculture and food issued in february 2025 by the European Commission still under discussion with the European Parliament.

1.1.2 *The Struggle for Soil Fertility*

Soil fertility, the cornerstone of healthy and productive agricultural systems, faces significant challenges in today's world. Organic production systems, however, offer solutions and align well with emerging agri-food policies that prioritize soil health.

Decades of intensive agriculture, characterized by heavy reliance on chemical fertilizers and monoculture practices, have led to nutrient depletion in many soils (FAO Soils Portal). This decline in essential nutrients like nitrogen, phosphorus, and potassium hinders plant growth and overall agricultural productivity.

Unsustainable land management practices, such as excessive tilling and deforestation, accelerate soil erosion. Wind and water carry away valuable topsoil, stripping the land of nutrients and organic matter, a crucial component for healthy soil structure and water retention (FAO Soils Portal).

In arid and semi-arid regions, irrigation practices can lead to salinization, where salts accumulate in the soil. This buildup creates a toxic environment for plants, rendering them infertile and unproductive (World Wildlife Fund).

Recognizing the gravity of these challenges, many agri-food policies are shifting towards promoting soil health and sustainability. Here are some key policy areas:

- **Supporting Organic Practices:** Policies that incentivize organic farming practices, such as crop rotation, cover cropping, and organic amendments, are gaining traction. These methods help replenish nutrients, improve soil structure, and promote microbial activity, leading to healthier and more fertile soils (International Federation of Organic Agriculture Movements (IFOAM)).
- **Reducing Chemical Inputs:** Policies aimed at reducing reliance on chemical fertilizers and pesticides are being implemented in many regions. This can help minimize soil degradation and promote the use of more sustainable alternatives, such as compost and biofertilizers (The European Commission).
- **Conservation Agriculture:** Policies that encourage conservation agriculture practices, like no-till farming and mulching, are becoming increasingly prevalent. These methods minimize soil disturbance, promoting the retention of organic matter and moisture, ultimately enhancing soil fertility (Food and Agriculture Organization of the United Nations (FAO)).

Organic production systems are at the forefront of soil fertility restoration. Here are some key strategies employed:

- **Nutrient Cycling:** Organic practices, like composting and manure application, promote nutrient cycling within the farming system. This reduces dependence on external inputs and helps maintain natural soil fertility (International Federation of Organic Agriculture Movements (IFOAM)).
- **Biodiversity Enhancement:** Organic farming encourages biodiversity by fostering healthy soil ecosystems with a diverse range of microorganisms. This biodiversity improves nutrient cycling, decomposition, and overall soil health (International Federation of Organic Agriculture Movements (IFOAM)).

- **Minimizing Soil Disturbance:** By minimizing tillage and promoting no-till methods, organic agriculture protects soil structure, prevents erosion, and promotes the growth of beneficial soil biota, all of which contribute to long-term soil fertility (International Federation of Organic Agriculture Movements (IFOAM)).

Soil fertility challenges demand innovative solutions and collaborative efforts. Agri-food policies that promote organic practices and prioritize soil health align perfectly with the strategies employed in organic production systems. By supporting these approaches, we can foster a future where agriculture not only sustains us but also nourishes the very foundation of our food production—the soil.

1.1.3 Organic Farming: The Soil, the Main Resource

At the heart of organic farming and livestock lies a fundamental element: the soil. More than just a base for plants, soil is a vibrant and complex ecosystem that sustains life in all its forms. Its health and vitality are essential pillars for successful and sustainable organic production.

Fertile soil isn't just a handful of dirt. It's a teeming universe with living organisms, from microscopic bacteria to earthworms, all working together to create a nutrient-rich environment conducive to plant growth. This biological diversity is crucial for decomposing organic matter, fixing nitrogen, retaining water, and structuring the soil—all vital processes for agricultural productivity (FAO Soils Portal).

Despite its paramount importance, soil is a finite and non-renewable resource. It takes thousands of years for fertile soil to form, while degradation can occur in decades through unsustainable agricultural practices like overuse of chemical fertilizers, intensive tilling, and erosion (FAO Soils Portal).

Organic Agriculture: A Regenerative Approach:

Organic farming and livestock recognize the vital importance of soil and adopt practices that promote its long-term health and conservation. These practices include:

- **Crop Rotation:** Alternating different types of crops helps prevent nutrient depletion and the proliferation of pests and diseases (Sustainable Review).
- **No-Till Farming:** Minimizing soil disturbance reduces erosion and preserves soil structure (USDA MyPlate).
- **Organic Fertilizers:** The incorporation of organic matter, like compost and manure, enriches the soil and boosts its natural fertility (Sustainable Review).
- **Integrated Pest and Disease Management:** Natural and biological control methods are prioritized, avoiding chemical pesticides that can harm soil life (Sustainable Review).

Organic Livestock: Rotational Grazing and Animal Welfare:

In organic livestock production, soil also plays a crucial role. Rotational grazing, which involves rotating livestock between different grazing areas, allows the soil to recover and stay fertile. Additionally, animal welfare practices, like access to natural pastures and shade, contribute to a healthy and sustainable pasture ecosystem (Rodale Institute).

A Sustainable Future:

By protecting and restoring soil health, organic farming and livestock not only guarantee the production of nutritious and high-quality food, but also contribute to environmental preservation and climate change mitigation. Healthy soil is essential for a sustainable future, both for agriculture and humanity as a whole.

1.1.4 Reducing Pesticides and Antibiotics for a healthier Future

The widespread use of pesticides and antibiotics in conventional agriculture poses significant health risks and environmental challenges. Organic production systems, on the other hand, prioritize alternative methods, aligning well with emerging agri-food policies that emphasize public health and environmental sustainability.

Challenges of Pesticides and Antibiotics:

- **Human Health Risks:** Exposure to pesticides through contaminated food and water can lead to a range of health problems, including neurological disorders, respiratory issues, and even certain cancers (World Health Organization). The overuse of antibiotics in livestock farming contributes to the rise of antibiotic-resistant bacteria, posing a serious threat to human health as common infections become resistant to treatment (Centers for Disease Control and Prevention).
- **Environmental Contamination:** Pesticides can contaminate soil, water resources, and non-target organisms, disrupting ecosystems and harming beneficial insects like pollinators (FAO Soils Portal). The overuse of antibiotics in livestock can also lead to the presence of antibiotic residues in manure and subsequently in soil and water, further amplifying the problem of antibiotic resistance (European Commission).
- **Reduced Biodiversity:** Reliance on pesticides often simplifies agricultural landscapes, leading to a decline in biodiversity. This reduction in plant and animal species can disrupt natural pest control mechanisms and weaken ecosystem resilience (The Rodale Institute).

Agri-Food Policies and Reduction Strategies:

Recognizing these challenges, agri-food policies are increasingly focused on promoting the responsible use of pesticides and antibiotics:

- **Integrated Pest Management (IPM):** Policies are encouraging the adoption of IPM strategies in agriculture. IPM relies on a combination of non-chemical methods like crop rotation, beneficial insects, and biological controls to manage pest populations, reducing reliance on synthetic pesticides (Food and Agriculture Organization of the United Nations (FAO)).

- **Antibiotic Stewardship Programs:** Many regions are implementing programs that promote responsible antibiotic use in livestock production. These programs emphasize preventative measures like improved hygiene and vaccination protocols, and encourage the use of antibiotics only when absolutely necessary (World Health Organization).
- **Supporting Organic Practices:** Policies that incentivize organic farming methods are gaining traction. Organic production strictly prohibits the use of synthetic pesticides and antibiotics, promoting safer food production and fostering healthier ecosystems (International Federation of Organic Agriculture Movements (IFOAM)).

Impact on Health:

By reducing reliance on pesticides and antibiotics, organic production and supportive agri-food policies can contribute to a healthier future in several ways:

- **Reduced Risk of Human Exposure:** Lower usage of pesticides translates to less contamination of food and water, minimizing the risk of associated health problems in consumers (The Organic Center).
- **Combating Antibiotic Resistance:** Responsible antibiotic use in livestock, as promoted by organic production and related policies, helps preserve the effectiveness of these life-saving drugs for human health (World Health Organization). The animal welfare is a principle in the organic livestock management which avoid the diseases and its treatments too.
- **Promoting Biodiversity:** Organic practices encourage biodiversity, which can create natural pest control mechanisms and support a healthier environment overall, leading to potential long-term health benefits for humans.

Addressing the challenges of pesticide and antibiotic overuse requires a multifaceted approach. Agri-food policies that prioritize these concerns, coupled with the adoption of organic production methods, can pave the way for a more sustainable and healthier future. By minimizing reliance on these harmful chemicals, we can protect human health, safeguard the environment, and ensure food production systems remain productive for generations to come.

1.2 Regulation

1.2.1 Origin in Europe

The 1970s witnessed a significant surge in the development and implementation of organic farming standards across several European countries. This period marked a crucial step forward for the organic movement, building upon the pioneering work done in Germany with the Biologische Wirtschaftsweise guidelines (1924) (Willer & Organic World 2023).

Several factors fueled the development of organic standards in Europe during this decade. The detrimental effects of intensive agriculture on soil health, water pollution, and biodiversity were becoming increasingly evident. Organic farming was seen as a potential solution to these environmental problems (European Commission 2023a). A growing awareness of the potential health risks associated with synthetic pesticides and fertilizers led to a rise in consumer demand for safe and healthy food. Organic products offered a credible alternative (European Commission 2023b). European governments began to recognize the potential benefits of organic farming and started offering support for the development of national organic standards (International Federation of Organic Agriculture Movements (IFOAM) 2023a).

The specific approaches to organic standards varied throughout Europe:

- **France:** In 1973, France established a system for the certification of "agriculture biologique" (biological agriculture) with a focus on soil health and the use of natural composting methods (Agence Bio 2023).
- **Switzerland:** The Swiss Association for Organic Farming (Sorganic) launched its certification program "Bio Suisse" in 1971, emphasizing animal welfare and closed-loop production systems (Bio Suisse 2023).
- **United Kingdom:** The UK Soil Association introduced its organic certification scheme in 1976, focusing on a holistic approach to sustainable farming practices (Soil Association 2023).
- **Other European Nations:** Several other European countries, including Denmark, Sweden, and Austria, also established their own organic farming standards during the 1970s (European Commission 2023c).

While the specific details differed between countries, these European organic standards generally shared some core principles:

- **Prohibition of Synthetic Inputs:** The use of synthetic pesticides, herbicides, and fertilizers was strictly prohibited.
- **Soil Fertility Management:** Organic soil amendments like compost and manure were emphasized to enhance soil health and productivity.
- **Natural Pest and Disease Control:** Farmers were encouraged to rely on natural methods like crop rotation, beneficial insects, and biological control agents.
- **Animal Welfare:** Higher welfare standards for livestock were often included, with requirements for access to pasture, organic feed, and humane treatment.

The development of organic farming standards in Europe during the 1970s had a profound impact on the global organic movement. These national programs provided a robust framework for organic production and certification, fostering consumer confidence and market growth. Additionally, the European standards served as a model for other countries around the world looking to establish their own organic regulations .

This period marked a critical turning point for organic farming. From a niche movement, it began transitioning into a more formalized and recognized system with the potential for broader adoption. The European standards paved the way for the development of international organic standards in the following decades.

1.2.2 The European Organic Regulation: A History of Continuous Improvement

European Organic Regulation was created in 1991: **"Council Regulation (EEC) N° 2092/91 of 24 June 1991 on organic production of agricultural products and indications referring thereto on agricultural products and foodstuffs"** also known as the **EU Organic Regulation 1**, on June 24th (FiBL Research Institute).

This initial regulation aimed to protect the use of terms like "organic", "ecological", "Biologic", "Bio" and "Eco" in relation to agricultural products sold within the EU. It established a set of production standards that farmers had to adhere to in order to label their products as organic. These standards focused on core principles like:

- **Prohibition of synthetic inputs:** The use of synthetic pesticides, herbicides, and fertilizers was strictly forbidden (Council Regulation).
- **Sustainable soil management:** Organic amendments like compost and manure were emphasized to maintain soil health and fertility (European Commission 2023d).
- **Animal welfare considerations:** While not initially a major focus, the regulation acknowledged the importance of animal welfare within organic production (FiBL Research Institute of Organic Agriculture).

A significant step forward came in 1994 when the regulation was amended to encompass livestock and animal products. This inclusion recognized the interconnectedness of organic agriculture and the need for a holistic approach. Livestock raised for organic production must now meet specific requirements, including:

- **Access to pasture and outdoor areas**
- **Organic feed**
- **Humane treatment** (Canadian Organic Regime)

This expansion ensured consistency across the entire organic food chain, from farm to table.

The rapid growth of the organic sector in the EU necessitated regular updates to the regulation. Between 1991 and 2009, the initial regulation was amended multiple times to address evolving needs and scientific advancements in organic farming practices (International Federation of Organic Agriculture Movements (IFOAM) 2023b).

On June 28th, 2007, a new and more comprehensive regulation, **Council Regulation (EC) No. 834/2007**, came into effect, replacing the original one. This updated regulation streamlined and strengthened the existing framework, encompassing all aspects of organic production and labeling within the EU.

The EU's commitment to continuous improvement in organic regulations culminated in the introduction of **Regulation (EU) 2018/848** on organic production and labelling of organic products on January first, 2022. This regulation, also known as the **EU Organic Regulation 2018**, incorporates the lessons learned from previous iterations and addresses emerging challenges. The intention of European Commission with this new update were:

- **Simplified procedures for certification**
- **Strengthened control systems to ensure consumer confidence**
- **Clearer rules for organic production methods and labeling**
- **Increased focus on sustainability and environmental protection** (World Wildlife Fund)

The development of the EU Organic Regulation serves as a testament to the EU's commitment to promoting sustainable and high-quality food production. This comprehensive framework has not only shaped the organic market within the EU but has also set the bar for organic standards globally, influencing regulations in other countries. As the organic sector continues to evolve, the EU can be expected to remain at the forefront, fostering innovation and ensuring the continued growth of a thriving organic food system.

1.3 The Certification System

1.3.1 European Producers' Certification: A Simplified Guide

In the bustling realm of European agriculture, organic products have carved a niche for themselves, promising a harmonious blend of sustainability and quality. To ensure the integrity of these products, a robust certification process safeguards their organic credentials. Let's delve into the intricacies of this process, guiding you through the steps involved in obtaining organic certification in Europe.

Step 1: Embracing the Organic Standards

The first step towards organic certification is to familiarize yourself with the comprehensive **EU Organic Regulation 2018/848**. This regulation outlines the meticulous standards that must be adhered to throughout the production process, encompassing aspects like:

- **Prohibition of Synthetic Inputs:** Synthetic pesticides, herbicides, and fertilizers are strictly forbidden, fostering a reliance on natural methods.

- **Sustainable Soil Management:** Organic soil amendments like compost and manure take center stage, nurturing soil health and fertility.
- **Animal Welfare:** Livestock raised for organic production must enjoy access to pasture, outdoor areas, and organic feed, ensuring humane treatment.

Step 2: Selecting a Control Body

Next, you'll need to select an **accredited control body**, an independent organization authorized to assess your compliance with the organic standards. These bodies conduct rigorous inspections and audits to verify your adherence to the regulations.

Step 3: Submitting an Application

With a chosen control body in hand, you'll embark on the application process. This typically involves providing detailed information about your farming practices, including:

- **Land Management:** A comprehensive overview of your land, including its history, crop rotation plans, and soil management practices.
- **Livestock Management:** Detailed plans for animal care, including feeding practices, access to outdoor areas, and veterinary treatments.
- **Traceability Systems:** A robust system to track the movement of organic products from farm to fork, ensuring transparency.

Step 4: Inspection and Audit

Once your application is deemed complete, the control body will schedule an inspection of your premises. This involves a thorough examination of your farming practices, including:

- **Field Inspection:** Assessing the condition of your land, crops, and livestock, ensuring compliance with organic standards.
- **Record Review:** Scrutinizing your records to verify the traceability of your products and adherence to regulations.

Step 5: Certification Decision

Following the inspection, the control body will evaluate their findings and make a certification decision. If your practices meet the organic standards, you'll receive an organic certificate, allowing you to label your products as organic.

Organic certification is not a one-time achievement but rather an ongoing commitment. Regular inspections by the control body ensure that your practices remain compliant with the organic standards, safeguarding the integrity of your products.

Organic certification opens doors to a growing market of consumers seeking sustainable and high-quality products. It not only enhances your brand reputation but also contributes to a healthier planet and promotes animal welfare. Embark on this journey of organic certification and reap the rewards of cultivating a harmonious blend of sustainability and quality.

1.3.2 European Certification a Guarantee System: Accreditation and Supervision

The **EU Organic Regulation 2018/848** establishes a comprehensive framework for organic production and labeling within the European Union (EU). This regulation includes a robust accreditation and supervision system to ensure that control bodies comply with the requirements of the regulation.

Accreditation is the process by which an independent body recognized by the **European Accreditation Cooperation (EA)**, assesses and verifies that a control body is competent to carry out inspections and certification of organic products. In order to be accredited, a control body must meet a number of criteria, including:

- Having qualified staff with expertise in organic production
- Having a transparent and documented inspection and certification procedure
- Having a system for ensuring the impartiality of its inspections and certifications

Supervision of control bodies is the ongoing monitoring of these bodies by the **European Commission** to ensure that they are complying with the requirements of the EU Organic Regulation 2018. The European Commission carries out the supervision of control bodies through a number of mechanisms, including:

1. *On-site inspections of control bodies*

   The European Commission conducts on-site inspections of control bodies to assess their compliance with the requirements of the regulation. These inspections typically involve:

   - Reviewing the control body's documentation, including inspection reports, certification records, and staff training records
   - Interviewing control body staff
   - Observing the control body's inspection procedures

2. *Reviews of control bodies' documentation*

   The European Commission also reviews control bodies' documentation on a regular basis to ensure that it is up-to-date and accurate. This documentation includes:

   - Inspection reports
   - Certification records
   - Staff training records
   - Complaints procedures

3. *Meetings with control bodies*

   The European Commission meets with control bodies on a regular basis to discuss any issues that have been identified during inspections or reviews of documentation. These meetings also provide an opportunity for the European Commission to provide guidance to control bodies on the implementation of the regulation.

4. *Investigations into complaints about control bodies*

   The European Commission investigates complaints about control bodies to ensure that they are being handled fairly and effectively. If a complaint is substantiated, the European Commission may take enforcement action against the control body, such as suspending or withdrawing its accreditation.

The European Commission has a number of mechanisms in place to ensure that control authorities carry out supervision of control bodies effectively. These mechanisms include:

- Issuing guidelines on the supervision of control bodies
- Providing training to control authorities
- Conducting audits of control authorities
- Taking enforcement action against control authorities that fail to comply with their obligations

The accreditation and supervision system is an important part of ensuring the integrity of the EU organic market. It helps to ensure that:

- Control bodies are competent to carry out inspections and certifications of organic products
- Control bodies are complying with the requirements of the EU Organic Regulation 2018
- Consumers can be confident that organic products have been certified by a competent and independent body

The accreditation and supervision system is a key component of the EU Organic Regulation 2018. It helps to ensure that the EU organic market is a fair and transparent market that operates in the best interests of consumers and the environment.

1.3.3 European Organic Certification: Access to International Markets

The European Union (EU) has established a comprehensive set of regulations for organic production and certification, ensuring the production of high-quality organic products. The EU organic logo is recognized and trusted worldwide, serving as a guarantee of compliance with strict organic standards.

The EU organic logo is widely accepted in numerous countries, facilitating trade of EU organic products to global markets. Some of the key markets where EU organic certification is sufficient include:

North America:

- **Canada:** The Canada-EU Comprehensive Economic and Trade Agreement (CETA) eliminates tariffs on most agricultural products, including organic goods, and recognizes the EU organic logo. **Agreement Date:** September 21, 2017.

- **United States:** While there is no formal trade agreement explicitly recognizing the equivalence of organic certifications, the United States Department of Agriculture (USDA) has determined that EU organic certification meets the standards of the USDA Organic Program (NOP) in most cases. Organic products exported from the EU to the US can carry the EU organic logo but must also comply with NOP labelling requirements to use the "USDA Organic" label.

Asia:

- **Japan:** The EU-Japan Economic Partnership Agreement (EPA) facilitates trade in organic products and recognizes the EU organic logo as equivalent to the Japanese Agricultural Standard (JAS) for organic products. Nevertheless buyers would ask for JAS Organic certification to accept products **Agreement Date:** February 1, 2019. https://trade.ec.europa.eu/access-to-markets/en/content/eu-japan-economic-partnership-agreement
- **South Korea:** The EU-South Korea Free Trade Agreement (FTA) streamlines trade in organic products and recognizes the EU organic logo as equivalent to the Korean Organic Certification Program (KOCP). **Agreement Date:** June 1, 2011. https://es.wikipedia.org/wiki/Tratado_de_Libre_Comercio_entre_Corea_del_Sur_y_Uni%C3%B3n_Europea

Latin America:

- **Mexico:** The EU-Mexico Free Trade Agreement (FTA) eliminates tariffs on most agricultural products, including organic goods, and recognizes the EU organic logo. **Agreement Date:** September 1, 2000. https://www.gob.mx/tlcuem
- **Chile:** The EU-Chile Association Agreement facilitates trade in organic products and recognizes the EU organic logo as equivalent to the Chilean Organic Production Standard (SCOC). **Agreement Date:** March 2, 2002. https://trade.ec.europa.eu/access-to-markets/en/content/eu-chile-association-agreement

Europe:

- **Switzerland:** Switzerland has a well-established organic market and is a key importer of EU organic products, especially dairy, meat, and fruit. In some supermarkets might be needed additional private labels to be accepted.
- **Norway:** Norway has a high consumption of organic products and imports a significant amount of EU organic products, including fruits, vegetables, and grains.

It is important to note that some of the markets mentioned above have established formal recognition agreements with the EU for organic certification. These agreements provide greater assurance to EU exporters by guaranteeing that their products will be accepted without the need for additional certifications.

Dates of Recognition Agreements:

- **Canada:** September 21, 2017 (CETA)
- **Japan:** February 1, 2019 (EPA)
- **South Korea:** June 1, 2011 (FTA)
- **Chile:** March 2, 2002 (Association Agreement)

Additional Considerations for Specific Markets:

Israel: An EU-Israel trade agreement includes provisions on the recognition of organic certification equivalence. Organic products exported from the EU to Israel can carry the EU organic logo and do not require additional labelling.
United Kingdon: Following the UK's withdrawal from the European Union, there were concerns about the continued recognition of EU organic certification in the UK. However, the UK-EU Trade and Cooperation Agreement (TCA), which came into effect on 1 May 2021, includes provisions for the continued recognition of EU organic products.

Key Points:

- **EU organic certification remains recognized in the UK.**
- **Organic products exported from the UK to the EU must still comply with EU organic regulations.**
- **There are a few exceptions to the mutual recognition of organic certification.**

Agreement Date: December 2023
Exceptions:

- **Certain processed products:** Some processed organic products, such as those containing non-organic ingredients, may require additional checks to ensure compliance with UK organic standards.
- **Pre-packaged food:** Pre-packaged organic food must be labelled in accordance with UK organic labelling requirements.

Additional Resources:

- **UK government guidance on organic food standards:** https://www.soilassociation.org/media/23380/gb-food-drink.pdf
- **European Commission information on organic trade agreements:** https://agridata.ec.europa.eu/Qlik_Downloads/Organic-Production-sources.htm

It is important for exporters of organic products to stay up-to-date on the latest requirements for organic certification in both the UK and the EU. Consulting with relevant authorities and trade organizations can provide valuable guidance and ensure compliance.

1.4 The Significance of Organic Production Associations: Ecovalia's Pioneering Role in Spain

Organic production associations have become **pivotal players** across the entire organic value chain, fostering sustainable agriculture. These organizations establish **rigorous standards** for organic production, ensuring product integrity. They also champion the cause by providing crucial **support to producers**, offering guidance on certification, technical assistance, and market access. Additionally, they play a vital role in **educating consumers**, empowering them to make informed choices.

Ecovalia stands as a shining example, spearheading the growth of the organic sector in Spain. Boasting a **mass social base of over 17,000** members, the association has earned a reputation as a leading authority on organic certification. Their pioneering efforts include establishing rigorous standards that have gained international recognition. Ecovalia's commitment extends to comprehensive support for producers through its representativeness, technical assistance, training programs, and market access opportunities. They actively engage in consumer education campaigns, raising awareness about the benefits of organic products and promoting informed purchasing decisions.

As consumer demand for organic products continues to surge, these associations will undoubtedly remain at the forefront of this transformative movement in sustainable agriculture.

References

Agence Bio (2023) Agriculture biologique. https://www.agencebio.org/

Bio Suisse (2023) Bio Suisse. https://www.bio-suisse.ch/fr/notre-association/portrait.html

Canadian Organic Regime. https://inspection.canada.ca/food-labels/organic-products/operating-manual/eng/1389199079075/1554143470958?chap=2

Centers for Disease Control and Prevention: Antibiotic Resistance Threats in the United States. https://www.cdc.gov/drugresistance/biggest-threats.html

Council Regulation (EEC) No 2092/91 of 24 June 1991 on organic production of agricultural products and indications referring thereto on agricultural products and foodstuffs. https://www.fao.org/faolex/results/details/en/c/LEX-FAOC002355/

European Commission (2023a) European green deal. https://commission.europa.eu/strategy-and-policy/priorities-2019-2024/european-green-deal_en

European Commission (2023b) Farm to fork strategy. https://ec.europa.eu/food/farm2fork_en

European Commission (2023c) Regulation (EU) 2018/848 on organic production and labelling of organic products. https://eur-lex.europa.eu/legal-content/EN/TXT/?uri=CELEX:32018R0848

European Commission (2023d) Regulation (EU) 2018/848 on organic production and labelling of organic products. https://eur-lex.europa.eu/eli/reg/2018/848/oj

European Commission: Farm to fork strategy. https://food.ec.europa.eu/system/files/2020-05/f2f_action-plan_2020_strategy-info_en.pdf

European Commission-Biodiversity Strategy. https://environment.ec.europa.eu/strategy/biodiversity-strategy-2030_en

European Commission-Farm to fork Strategy. https://food.ec.europa.eu/horizontal-topics/farm-fork-strategy_en

FAO Soils Portal
FAO Soils Portal. https://www.fao.org/soils-portal/soil-degradation-restoration/en/
FAO Soils Portal. https://www.fao.org/soils-portal/en/
FiBL & IFOAM – Organics International (2024) The world of organic agriculture. https://www.fibl.org/en/shop-en/1747-organic-world-2024
FiBL Research Institute
FiBL Research Institute of Organic Agriculture. https://www.fibl.org/en/
Food and Agriculture Organization of the United Nations (FAO): Conservation Agriculture. https://www.fao.org/conservation-agriculture/en/
Food and Agriculture Organization of the United Nations (FAO): Integrated Pest Management. https://www.fao.org/agriculture/crops/thematic-sitemap/theme/spi/scpi-home/managing-ecosystems/integrated-pest-management/en/
Health Canada – Canada's Food Guide. https://food-guide.canada.ca/en/
International Federation of Organic Agriculture Movements (IFOAM) (2023a) About IFOAM. https://www.ifoam.bio/about-us
International Federation of Organic Agriculture Movements (IFOAM) (2023b) https://www.ifoam.bio/
International Federation of Organic Agriculture Movements (IFOAM): The Organic Standard. https://www.ifoam.bio/our-work/how/standards-certification/organic-guarantee-system/ifoam-standard
International Federation of Organic Agriculture Movements (IFOAM): The Organic Standard. https://gemini.google.com/%5B%3C1%3E, https://www.ifoam.bio/our-work/how/standards-certification/organic-guarantee-system/ifoam-standard%5D, https://www.ifoam.bio/our-work/how/standards-certification/organic-guarantee-system/ifoam-standard
Rodale Institute: Why Organic? Organic Basics: Regenerative Organic Agriculture. https://rodaleinstitute.org/why-organic/organic-basics/regenerative-organic-agriculture/
Soil Association (2023) Soil association. https://www.soilassociation.org/
Sustainable Review: Organic vs. Sustainable Farming. https://sustainablereview.com/organic-vs-sustainable-farming/
The European Commission: Farm to fork strategy. https://food.ec.europa.eu/horizontal-topics/farm-fork-strategy_en
The Organic Center: Benefits of Organic Food. https://www.organic-center.org/our-work
The Rodale Institute: Why Organic? Organic Basics: Biodiversity. https://rodaleinstitute.org/why-organic/organic-basics/
USDA MyPlate. https://www.choosemyplate.gov
USDA National Organic Program. https://www.ams.usda.gov/about-ams/programs-offices/national-organic-program
Willer H, Organic World (2023) The world of organic agriculture 2023. FiBL Research Institute of Organic Agriculture & IFOAM – Organics International
World Health Organization: Pesticides. https://www.who.int/news-room/fact-sheets/detail/pesticides
World Health Organization: WHO guidelines on the prudent use of antimicrobials in animals. https://aricjournal.biomedcentral.com/articles/10.1186/s13756-017-0294-9
World Health Organization: Antibiotic Resistance. https://www.who.int/antimicrobial-resistance/en/
World Wildlife Fund: Soil Salinization. https://www.worldwildlife.org/threats/soil-erosion-and-degradation

Chapter 2
European Organic Food Markets and Consumer Preferences

Katrin Zander

Abstract The markets for organic food are constantly growing in Europe. This is on the one hand due to political goals of increasing organic farming area due to overall sustainability goals and, on the other hand to marked consumer interest in organic food. Main reasons for consumers to buy organic are environmental and health reasons, most important barriers are high prices for organic food, low consumer knowledge and trust. Increasing organic markets needs sound and well targeted communication with consumers.

Keywords Consumer preferences · Intention-behaviour gap · Marketing of organic product · Organic food · Organic markets · Purchase behaviour

2.1 Introduction

Organic farming ranks high on the EU Farm-to-Fork (F2F) strategy which is at the heart of the European Green Deal. It aims at redesigning our food system in order to become more sustainable. According to the F2F strategy 25% of agricultural land should be farmed organically by 2030. Main driver for this aim are sustainability and the environmental benefits of organic farming which have been repeatedly documented (e.g. Sanders and Hess 2019). There might be other forms of sustainable farming throughout Europe, the very big advantage of organic is that clear standards have been set by the EU commission with the Regulation (EU) 2018/848. The first common EU regulation was released already in 1991 and, well protected by this EU regulation on organic farming, an increasing market for organic food has developed.

Given the EU target of increasing organic area, the challenge remains to identify effective tools to increase supply and demand. The willingness of producers to start and/or to increase organic production highly depends on the economic viability

K. Zander (✉)
University of Kassel, Witzenhausen, Germany
e-mail: k.zander@uni-kassel.de

C. Palacios Riocerezo, M. Gonzalez Ronquillo (eds.), *Traditional Livestock Production*, https://doi.org/10.1007/978-3-031-88275-3_2

which might be granted by subsidies or by the market. In almost all EU member states governments are offering subsidies to farmers, often in form of area payments for organic acreage. To rely only on producer subsidies bears the risk of increasing supply without increasing demand accordingly, which probably will result in decreasing market prices. That is why any policy aiming at increasing organic production in the EU also needs to have in mind the demand side and the market interactions which depend to a large degree on consumers' purchase decisions. But not only at the larger policy level knowledge about consumer preferences and behaviour is important, also at individual producer level this kind of knowledge might be decisive for success.

This book chapter will shortly describe the history of organic markets within the EU during the last 30 years, with an emphasis on the most recent developments shaped by COVID-19 measurements and the Ukraine war. Consumer behaviour for organic (and other sustainable) food will be explained, as well as reasons to buy and not to buy organic food. The so-called attitude behaviour gap which frequently can be observed when it comes to sustainable food consumption will be addressed as well as approaches to reduce this gap. This leads to the identification of the relevant target group for the very specific product at hand, if it is organic or any other sustainably produced food, and of the development of well targeted communication strategies.

2.2 Organic Markets in Europe

The markets for organic food are continuously growing aver the last 20 years and more and more consumers are buying organically produced food (Fig. 2.1). Like all food markets, the organic food markets have been highly influenced by the COVID-19 pandemic. Due to restrictions in out of home consumptions people had to provide themselves with food in the different retail shops where they could opt for the organic alternative. In out of home consumption the organic alternative is still rare. This is one of the main reasons for the marked increase in organic sales on average of about 15% (Europe) and 8% (EU) between 2019 and 2020. Some countries faced much higher market increases such as Germany (22.3%), Switzerland (19.1%) and Austria (18.2%). In 2021 retail sales further increased on average by at about 3.7% (Willer et al. 2023) and a total market volume of at 46.7 bill € (EU) and 54.5 bill € (Europe) was reached. A slight decrease in European retail sales has to be stated for 2022 (53.1 bill € in Europe and 45.1. bill € in the EU). Consumers were challenged due to increasing prices, caused by high inflation rates due to the trade distortions caused by the pandemic and the following Russian Ukraine war. Another important, sometimes neglected, reason for decreasing organic sales is the restart of out-of-home consumption in 2022, also due to recreational activities and travelling. More recent numbers exist for Germany for 2023 and show, following the decline in organic sales in 2022, a slight increase of 5% which, however, is mainly driven by increasing prices.

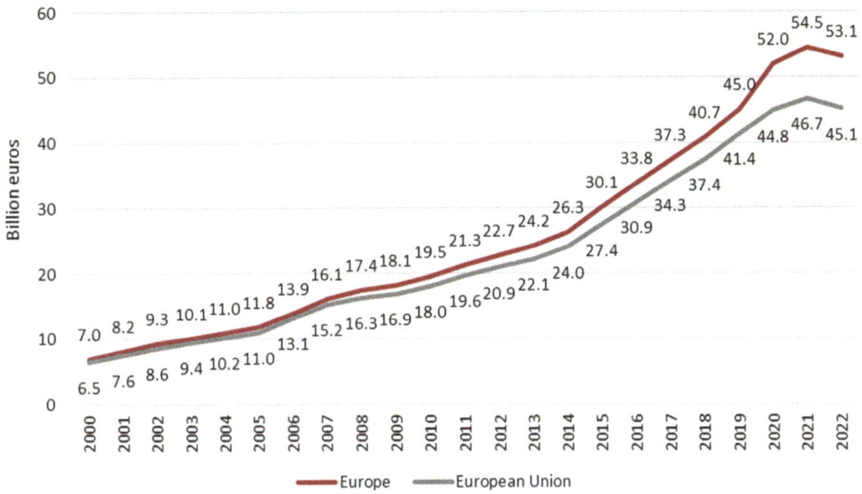

Fig. 2.1 Europe and European Union: Development of retail sales of organic food (2000 to 2022). (Source: Own compilation based on Willer et al. (various years))

Although these numbers for the retail sales seem to be quite impressive, when comparing with overall turnover of the food sector, the relevance of the organic sector is low. The share of organic in all food sales was at about 13% in Denmark, 11.6% in Austria, 10.9% in Switzerland and 7% in Germany in 2021. Generally spoken organic market shares are higher in northern European countries than in southern European countries. This also reflects in the per capita spending for organic food: it is highest in some mid to northern European countries, such as Switzerland (425€), Denmark (384€) as well as Sweden (262€). Large producing countries of organic food in the South, like Italy and Spain show marked lower numbers (Italy 67€, Spain 53€). Eastern European countries tend to have even lower per capita spendings on organic food (AMI 2023).

Looking at product groups with animal origin, eggs and dairy products have rather high market shares, while meat usually goes along with low market shares. Organic eggs and dairy products are more important than any other organic products, except fruit and vegetables. In some countries, such as Denmark and France organic eggs have monetary market shares of about 30%. For dairy products the market shares are at about 10% in Sweden and in Switzerland. The market shares of meat are markedly lower in general, at 2.9% in Sweden and 6.2% in Switzerland. Some biases have to be considered since organic meat usually is much more expensive than conventional meat so that the physical organic share is even smaller. Main reason for lower organic market shares for meat compared to the overall organic market share are the high price premia for organic products since conventional meat is very cheap (Willer et al. 2023). Another important reason is that organic consumers tend to eat less meat since they are more aware of the climate impact of meat consumption, of the high resource use as well as of adverse health effects of meat-based diets (s.a. Apostolidis and McLeay 2019; Sigrist and Hartmann 2019).

2.3 Reasons to Buy Organic Food

When planning to sell organic food to consumers and to get product prices which cover the costs of organic production an important starting point is to think about the target group. The target group is that part of consumers which might be interested in the organic offer. The same applies when going into organic specialties, e.g. very specific meat products from small-scale husbandry. It is necessary to know the specific needs and preferences of the target group. As a starting point we try to better understand why an increasing share of consumers buy organic food.

An important characteristic of consumers in affluent Western societies is that not only material product properties such as ingredients, nutritional value, flavour or freshness, but also additional, non-material characteristics of the production process are decisive for consumers' purchasing behaviour. A growing share of consumers are dissatisfied with standardised and anonymous foods that are produced under unknown environmental and social conditions. They are asking for more information, greater traceability and transparency about product origins and production processes (Grymshi et al. 2021; Li and Kallas 2021; Fuller and Grebitus 2023). Purchasing behaviour in which consumers base their decision making on the existence of ethical product or process properties, such as fair trade, local origin, traditionally produced, higher animal welfare or organic production, is referred to as ethical consumption. The ethical product or process characteristics can mainly be categorised according to the three pillars of sustainability: ecology, economy and social responsibility. A forth category referring to a wider understanding of further ethical issues including animal welfare might complement these three pillars (Carrigan 2018; Michelsen 2014; Zander et al. 2013).

Organic food melds several of these sustainable or ethical attributes, such as being environmentally friendly by largely avoiding the use of synthetic chemical pesticides, easily soluble fertilisers and genetically modified organisms. Consumers particularly appreciate higher animal welfare in husbandry and the naturalness of the products. Naturalness of organic products is related to environmental friendliness and animal welfare as well as to health and sensory perception (Hemmerling et al. 2016). Due to lower pesticide residues and lower degrees of processing organic food is perceived to be healthier (Apaolaza et al. 2018). Research shows that the motives for buying organic food are generally divided into hedonistic or selfish (taste, health) and altruistic motives (environmental protection, animal welfare). The relative importance of hedonistic versus altruistic purchase motives vary largely between individual consumers (Aertsens et al. 2009; Schleenbecker and Hamm 2013; Meyer-Höfer et al. 2014; Hemmerling et al. 2016; Zander et al. 2018).

Research also shows that local production is perceived to be very important, although not being a fundamental characteristic of organic products (Zander et al. 2018; Schleenbecker and Hamm 2013).

2.4 Barriers to Organic Food Purchasing

Many studies elaborate on consumers' preferences for organic food, but when it comes to the act of purchasing they decide for the conventional alternative. The fact that (ethical) attitudes and even intentions are not directly translated into actions at the point of sale is referred to as "intention-behaviour gap"(also attitude-behaviour gap). This phenomenon can also be observed with organic food (Padel and Foster 2005; Vermeir and Verbeke 2006; Sultan et al. 2020; Aschemann-Witzel and Niebuhr Aagaard 2014; Schaeufele and Hamm 2018). The intention-behaviour gap has been intensely studied but cannot be fully explained (Sultan et al. 2020), and better understanding the intention-behaviour gap is of crucial relevance in order to motivate consumers not only to think but also to act sustainably.

The existence of the attitude-behaviour gap or intention-behaviour gap is often referred to as 'bounded rationality' or irrational consumer behaviour and, sometimes, is used as a proof of the low validity of (scientific) consumer studies. However, there are reasons why observable behaviour may deviate from attitudes and intentions. Some of these reasons are addressed in the next sections: lack of knowledge, lack of trust, price differences and limited product availability.

Knowledge of how to behave ethically is an important prerequisite for transferring ethical attitudes into ethical purchase behaviour. Without the knowledge about how organic production processes and products may contribute to individual ethical attitudes, consumers cannot adapt their purchasing behaviour accordingly (Stanton and Cook 2019). Previous research has shown that consumer knowledge of organic standards and certification processes is often limited, even in the relatively well-established organic food markets like e.g. the German one (Janssen and Hamm 2012; Zander et al. 2015; Brümmer et al. 2019; Stanton and Cook 2019). Usually, there is often some basic understanding, for example, of the restricted use of pesticides. In addition, consumers have their own "intuitive" expectations of organic farming methods, e.g. with regard to local production or production in small structures (Zander et al. 2015; Brümmer et al. 2019). These "intuitive" quality expectations do not always correspond to reality, which may lead to misunderstandings and even rejection due to the individual perception to have been cheated.

Providing information is the most common way to combat a lack of knowledge. However, consumers are confronted with a huge amount of information in their daily lives. Whereas in the past the availability of information and the cost of obtaining information were limiting factors in the search for information, the task of today's consumers is to actively select the information relevant to them and for their individual decision-making, including identifying and disregarding irrelevant information. This means that they have to limit their individual information flow in order to make decisions in line with their attitudes. To do this, they use so-called "mental shortcuts", i.e. shortcuts or simplifications that result in decisions not being scrutinised in detail, but instead being made habitually or by searching for certain key information (Stanton and Cook 2019).

Credibility and trust are particularly important in sustainable or ethical market segments due to the pronounced "information asymmetry". "Information asymmetry" means that producers and consumers have very different levels of information. Producers know everything, at least about their own production process, while consumers have to trust the information given, e.g. about organic production. This information asymmetry is addressed by independent certification schemes, such as the ones following the EU regulations on organic farming. Nevertheless, lack of trust in standard setting, certification and labelling was often cited by consumers as a reason not to buy 'organic' (Sultan et al. 2020; Padel and Foster 2005; Brümmer and Zander 2020; Katt and Meixner 2020). This is remarkable because the production standards for organic products are well defined and certification systems work rather well. Another very important reason for not buying organic food are high prices (Rödiger et al. 2016; Brümmer and Zander 2020; Aschemann-Witzel and Zielke 2017). In most cases, organic food is more expensive than their conventional counterparts. European food markets to a large extent are characterised by fierce price competition, this is also the case for organic food markets. Consumers are used to compare prices and to look for cheaper options. That is why consumers' willingness to pay is perceived to be a key indicator for the market success of organic food. Both, observations on the food markets as well as scientific studies show that, at least part of, consumers are willing to pay price premia for organic food. This average willingness to pay vary greatly and depend, among others, on the product group (e.g. Hemmerling et al. 2015; Aschemann-Witzel and Zielke 2017). Usually, it is higher for fruit and vegetables, meat, cheese and bread, while it is lower for milk, cereals and potatoes (Van Loo et al. 2013; Rödiger and Hamm 2015; Hemmerling et al. 2015). Studies report an additional willingness to pay of around 22% for organic milk (Aldanondo-Ochoa and Almansa-Sáez 2009) and organic yoghurt (Van Loo et al. 2013 of 120% for organic eggs (Janssen and Hamm 2014) and of 16% (Brugarolas Mollá-Bauzá et al. 2005). Price relations between organic and conventional food might have changed and so do consumers' willingness to pay.

Some research challenges the general notion of price as purchase barrier for organic food. First, consumers and their willingness to pay vary widely and averages over all consumers neglects different preferences (Jensen et al. 2019). Second, consumers show a low price knowledge for many food products (Rödiger et al. 2016; Aschemann-Witzel and Zielke 2017), and third, prices may also serve as quality cue which means that more expensive products are perceived to be of higher quality and therefore preferred over the cheaper option (Goldstein et al. 2008; Mitra 1995).

Given the low price knowledge of many consumers when it comes to organic food, it is surprising that higher prices are often cited as a reason for not buying organic food. One possible explanation is that organic food has the image of being expensive (Hamm et al. 2007), so that the purchase of organic food is not even considered and therefore no price comparison takes place. Another reason might be that naming high prices as purchase barrier might be socially better accepted than lack of preferences for this kind of environmentally friendly and healthy food.

The high price image of organic food can therefore be a decisive obstacle, particularly for new or occasional organic customers. Price perception is also important in connection with the credibility of organic food. A lack of trust in organic certification systems, combined with the subjectively perceived risk of being cheated and not getting what is expected, undermines the willingness to pay the higher prices for organic food.

Finally, a limited availability of organic food may hamper consumption (Katt and Meixner 2020). Although the availability of organic food has greatly improved in recent decades, the breadth and depth of organic food on offer in many shopping centres does not match that of conventional food by far. This means that consumers might have to visit several shops to fill their shopping basket with the desired variety of organic products. This is in contrast to one-stop shopping preferred by many consumers.

Various studies have shown that a significant share of consumers prefer local products over organic products (Hempel and Hamm 2016), meaning that there is competition between organic and local foods. The reasons for the preference for local food lie in the lower environmental impact (food miles), the perception of greater trust and transparency, authenticity, smaller farms and firms and the desire to strengthen local structures (Brümmer et al. 2019; Jensen et al. 2019). The greater trust in local food is somehow surprising, since the term 'local' is neither clearly defined nor legally protected. Nevertheless, due to its positive perception by consumers it is widely used in the food retail. Obviously, the term 'local' seems to directly address the emotional level of consumers, and might be used as the above mentioned 'mental shortcut' in decision making.

2.5 'Organic' Consumer Segments

Consumers' buying organic food are very divers. There are those how buy only sometimes organic food and others who do it on a regular basis. Often, distinction is made between "occasional" and "regular" or "frequent" buyers, depending on the frequency or intensity of organic purchases. Sometimes a third intermediate group is also defined (Janssen and Hamm 2014; Van Loo et al. 2013). Hjelmar (2011) distinguishes between "politically/ethically" orientated consumers and "pragmatic" consumers, Mondelaers et al. (2009) between "heavy" and "light" consumers and Hempel and Hamm (2016) between "ecologically orientated" and "non-ecologically orientated" consumers. There is a strong overlap between "heavy", "regular" and "politically/ethically orientated" consumers on the one hand and "light", "occasional" and "pragmatic" consumers on the other hand. From a marketing perspective, it is important that consumers in the latter group easily switch to conventional products, particularly when their expectations of product quality and price are not met. At the same time, this group is an interesting target group when it comes to increasing demand, e.g. by emphasising the benefits of organic food and its credibility through well targeted communication (Rödiger et al. 2016).

Better knowing the common properties of members of the different groups is precondition for successful marketing of organic food. Potential customers need to be effectively addressed through targeted marketing activities such as product design, pricing, distribution and communication. Regular organic food consumers are described as being better educated, middle-aged, wealthier, more often female, mostly with children (e.g. Hughner et al. 2007; Aertsens et al. 2009; Hjelmar 2011) and better informed about organic food (Vermeir and Verbeke 2006). More than the socio-demographic characteristics, regular organic consumers differ in their attitudes, behavioural characteristics and beliefs. Organic consumers are characterised by a stronger concern for the natural environment, for animal welfare and for local production as well as a strong interest in pro-social behaviour. They are convinced of the higher quality and better taste of organic food (Hjelmar 2011).

2.6 Perspectives for Organic Food Markets

When it comes to the marketing of organic meat another challenge has to be considered. As mentioned before organic consumer consume less meat. In addition, they are more concerned about the effect of their individual consumption on the environment which is in fact an important purchase reason for organic food. Meat (and dairy) products are aligned with higher carbon footprints than plant-based alternatives. This means that motivating consumers to buy organic meat can be challenging and needs a well elaborated communication strategy focussing on the specific advantages of animal husbandry, e.g. for biodiversity or the use of permanent grassland.

Consumer demand for organic food varies largely between European countries and between individuals, and so do motives for buying and for not buying organic food. While on the one hand this might seem to be discouraging, on the other hand it opens up large potential for successful marketing of organic product. Generally, organic food does not target the mass markets but specific consumer segments which appreciate the specific quality and which are willing to pay higher product prices. But, with the aim of increasing organic production up to 25% share of organic area, larger parts of consumers will have to be addressed.

References

Aertsens J, Verbeke W, Mondelaers K, van Huylenbroeck G (2009) Personal determinants of organic food consumption: a review. Br Food J 111:1140–1167

Aldanondo-Ochoa AM, Almansa-Sáez C (2009) The private provision of public environment: consumer preferences for organic production systems. Land Use Policy 26:669–682

AMI (2023) Marktbilanz Öko-Landbau 2023. Agrarmarkt Informations-Gesellschaft, Bonn
Apaolaza V, Hartmann P, D'Souza C, López CM (2018) Eat organic – feel good? The relationship between organic food consumption, health concern and subjective wellbeing. Food Qual Prefer 63:51–62
Apostolidis C, McLeay F (2019) To meat or not to meat? Comparing empowered meat consumers' and anti-consumers' preferences for sustainability labels. Food Qual Prefer 77:109–122
Aschemann-Witzel J, Niebuhr Aagaard EM (2014) Elaborating on the attitude–behaviour gap regarding organic products: young Danish consumers and in-store food. Int J Consum Stud 38(5):443–569
Aschemann-Witzel J, Zielke S (2017) Can't buy me green? A review of consumer perceptions of and behaviour toward the price of organic food. J Consum Aff 2017(Spring):211–251
Brugarolas Mollá-Bauzá M, Martínez-Carrasco Martínez L, Martínez Poveda A, Rico Pérez M (2005) Determination of the surplus that consumers are willing to pay for an organic wine. Span J Agric Res 3(1):43–51
Brümmer N, Zander K (2020) Drivers of organic food choice in Germany—the case of young adults. Org Agric 10:57–64. https://doi.org/10.1007/s13165-020-00306-2
Brümmer N, Klawitter M, Zander K (2019) Werthaltungen, Einstellungen und Präferenzen junger Erwachsener zum ökologischen Landbau und seinen Produkten. Abschlussbericht BLE-BÖLN Projekt 2815OE001, Braunschweig. https://orgprints.org/37784/
Carrigan M (2018) Food and ethical consumption. J Consum Ethics 2(2):1–7
Fuller K, Grebitus C (2023) Consumers' preferences and willingness to pay for coffee sustainability labels. Agribusiness 39(4):1007–1025. https://doi.org/10.1002/agr.21810
Goldstein R, Almenberg J, Dreber A, Emerson JW, Herschkowitsch A, Katz J (2008) Do more expensive wines taste better? Evidence from a large sample of blind tastings. J Wine Econ 3(1):1–9
Grymshi D, Crespo-Cebada E, Elghannam A, Mesías FJ, Díaz-Caro C (2021) Understanding consumer attitudes towards ecolabeled food products: a latent class analysis regarding their purchasing motivations. Agribusiness 38:93–107. https://doi.org/10.1002/agr.21714
Hamm U, Aschemann J, Riefer A (2007) Sind die hohen Preise für Bio-Lebensmittel wirklich das zentrale Problem für den Absatz? Berichte über Landwirtschaft 85(2):252–271
Hemmerling S, Hamm U, Spiller A (2015) Consumption behaviour regarding organic food from a marketing perspective – a literature review. Org Agric. Online first 5:277–313. https://doi.org/10.1007/s13165-015-0109-3
Hemmerling S, Asioli D, Spiller A (2016) Core organic taste: preferences for naturalness-related sensory attributes of organic food among European consumers. J Food Prod Mark 22(7):824–850. https://doi.org/10.1080/10454446.2015.1121428
Hempel C, Hamm U (2016) Local and/or organic: a study on consumer preferences for organic food and food from different origins. Int J Consum Stud 40(6):732–741
Hjelmar U (2011) Consumers' purchase of organic food products. A matter of convenience and reflexive practices. Appetite 56:336–344
Hughner RS, McDonagh P, Prothero A, Shultz CJ, Stanton J (2007) Who are organic food consumers? A compilation and review of why people purchase organic food. J Consum Behav 6(5):94–110
Janssen M, Hamm U (2012) The mandatory EU logo for organic food: consumer perceptions. Br Food J 114(3):335–352
Janssen M, Hamm U (2014) Governmental and private certification labels for organic food: consumer attitudes and preferences in Germany. Food Policy 49:437–448
Jensen JD, Christensen T, Denver S, Ditlevsen K, Lassen J, Teiuber R (2019) Heterogeneity in consumers‘ perceptions and demand for local (organic) food products. Food Qual Prefer 73:255–265
Katt F, Meixner O (2020) A systematic review of drivers influencing consumer willingness to pay for organic food. Trends Food Sci Technol 100:374–388
Li S, Kallas Z (2021) Meta-analysis of consumers' willingness to pay for sustainable food products. Appetite 163:105239

Meyer-Höfer M, Nitzko S, Spiller A (2014) Is there an expectation gap? Consumers' expectations towards organic. Br Food J 117(5):1527–1546
Michelsen G (2014) Education for sustainable development: status quo and perspectives. Int Yearb Res Arts Educ 2:121–129
Mitra A (1995) Price cue utilization in product evaluations. The moderating role of motivation and attribute information. J Bus Res 33:187–195
Mondelaers K, Verbeke W, Huylenbroeck G (2009) Importance of health and environment as quality traits in the buying decision of organic products. Br Food J 111(10):1120–1139
Padel S, Foster C (2005) Exploring the gap between attitudes and behaviour. Understanding why consumers buy or do not buy organic food. Br Food J 107(8):606–625
Rödiger M, Hamm U (2015) How are organic food prices affecting consumer behaviour? A review. Food Qual Prefer 43:10–20
Rödiger M, Plassmann S, Hamm U (2016) Organic consumers' price knowledge, willingness-to-pay and purchase decision. Br Food J 118(11):2732–2743
Sanders J, Hess J (eds) (2019) Leistungen des ökologischen Landbaus für Umwelt und Gesellschaft. 2. überarbeitete und ergänzte Auflage. Johann Heinrich von Thünen-Institut, Braunschweig, 398 p, Thünen Rep 65. https://doi.org/10.3220/REP1576488624000. Online at: https://www.thuenen.de/media/publikationen/thuenen-report/Thuenen_Report_65.pdf
Schaeufele I, Hamm U (2018) Organic wine purchase behaviour in Germany: exploring the attitude-behaviour gap with data from a household panel. Food Qual Prefer 63:1–11
Schleenbecker R, Hamm U (2013) Consumers' perception of organic product characteristics. A review. Appetite 71:420–429
Sigrist M, Hartmann M (2019) Impact of sustainability perception on consumption of organic meat and meat substitutes. Appetite 132:196–202
Stanton JV, Cook LA (2019) Product knowledge and information processing of organic foods. J Consum Mark 36(1):240–252
Sultan P, Tarafder T, Pearson D, Henrycks J (2020) Intention-behaviour gap and perceived behavioural control-behaviour gap in theory of planned behaviour: moderating roles of communication, satisfaction and trust in organic food consumption. Food Qual Prefer 81:103838
Van Loo EJ, Hoang Diem MN, Pieniak S, Verbeke W (2013) Consumer attitudes, knowledge, and consumption of organic yoghurt. J Dairy Sci 96:2118–2129
Vermeir I, Verbeke W (2006) Sustainable food consumption. Exploring the consumer "attitude-behavioral intention" gap. J Agric Environ Ethics 19(2):169–194
Willer H, Schlatter B, Travnicek J (eds) (2023) The world of organic agriculture. FIbL & IFOAM. Online at: http://organic-world.net/yearbook/yearbook-2023.html
Zander K, Stolz H, Hamm U (2013) Promising ethical arguments for product differentiation in the organic food sector. A mixed methods research approach. Appetite 62:133–142
Zander K, Padel S, Zanoli R (2015) EU organic logo and its perception by consumers. Br Food J 117(5):1506–1526
Zander K, Schleenbecker R, Hamm U (2018) Consumer behaviour in the organic and fairtrade food market in Europe. In: Parvathi P, Grote U, Waibel H (eds) Fair trade and organic agriculture: a winning combination? CAB International, Wallingford, pp 51–60

Chapter 3
Carbon Footprint in Organic Livestock Farming

Jaime Nieto de la Losa, Javier Plaza Martín, and Carlos Palacios Riocerezo

Abstract Greenhouse gases (GHG) levels in the atmosphere has been significantly increased, in the last few years, thereby contributing to the damaging effects caused by the global warming. Particularly, the average temperature on Earth in the past decade increased by 1.09 °C compared to the reference data from 1850 to 1900, according to the recent Intergovernmental Panel on Climate Change (IPCC) report. In this context, it is widely accepted that a considerable amount of the anthropogenic GHG emissions stems from highly intensified agricultural and livestock practices. This kind of models try to maximize their productive rates at any cost, using fewer natural resources and extremely-high stocking rates, therefore diminishing environmental sustainability and animal welfare. In 2014 the EU Commission stated that every EU member should reduce their GHG emissions by 30% by 2030 compared to the 2005 level, promoting in consequence the progressive disappearance of super intensive productive models in the primary sector. In addition, there is in the present society an increasing trend towards more sustainable productive and management activities, such as those promoted by extensive and/or organic systems. All this seems to point to a slow yet unstoppable transition towards more sustainable production systems, such as organic systems. In organic livestock systems, the main sources of GHG emissions are associated with enteric fermentation in ruminants' farms or with manure management for pig farms, but not with purchasing feed inputs or using chemical products. Several studies have shown that, a priori, extensive systems, standing out organic practices, have a lower environmental impact than intensive systems because they use their own inputs instead of external ones, thereby reducing their energy consumption and, consequently, the fossil fuels they require. Moreover, they could also generate added value from an economic and environmental point of view. Therefore, it could be assumed that organic farming will be crucial in the future food systems. According to the International Federation of Organic Agriculture Movements (IFOAM), in a foreseeable scenario

J. Nieto de la Losa (✉) · J. Plaza Martín · C. Palacios Riocerezo
Area of Animal Production, Faculty of Agricultural and Environmental Sciences, University of Salamanca, Salamanca, Spain
e-mail: jaimenl@usal.es; pmjavier@usal.es; carlospalacios@usal.es

C. Palacios Riocerezo, M. Gonzalez Ronquillo (eds.), *Traditional Livestock Production*, https://doi.org/10.1007/978-3-031-88275-3_3

where organic farming accounts for 50% of food production, it would yield a potential reduction of 12–14% in the GHG emissions from farming industry in the EU. Measuring the carbon footprint of livestock farming of extensive systems and specifically, of organic ones, is a major objective to sustain the current productive transition. However, GHG emissions derived from farming systems are considered quite complex and heterogeneous and, in addition, there is a wide variety of methods that can be used to calculate or estimate the carbon footprint. All this together make the measurement and comparison of GHG emissions from different geographical and cultural areas a challenging task. Among those methods, on of the most internationally-recognized is the well-known Life Cycle Assessment (LCA), which allows the measurement of GHG emissions both in conventional and organic productive systems. But also, the emission factors involved in this calculation will determine the carbon footprint obtained, as it is quite different consider that the limits of the system remain on the farm itself or, on the contrary, consider the entire life cycle of the inputs (transport, manufacturing, etc.).

The organic livestock farming model might be a feasible strategy for reducing GHGs and, therefore, for meeting the global food demand while maintaining environmental sustainability and animal welfare.

Keywords Carbon footprint · Carbon sequestration · Greenhouse gases (GHG) · Life cycle assessment (LCA) · Sustainability

3.1 Introduction

Greenhouse gas (GHG) levels in the atmosphere have markedly risen in recent years, exacerbating the adverse impacts of global warming. This indisputable warming trend cannot be solely attributed to natural climate variability. Human activities, notably the combustion of coal and oil, have precipitated a substantial uptick in atmospheric GHG concentrations, which trap heat. As human emissions of these gases persist, the Earth is anticipated to undergo further warming in the ensuing decades and centuries (VijayaVenkataRaman et al. 2012).

In particular, a report by the Intergovernmental Panel on Climate Change (IPCC) notes that global temperatures are currently the highest on record since 1885. These temperatures are projected to escalate further, reaching a range of approximately 6.4 °C by the year 2100 as a consequence of global warming (Ahmed et al. 2018).

Climate change is one of the outcomes of CO_2 emissions and global warming, precipitating enduring alterations in regional and global weather patterns. It encompasses shifts in temperature, precipitation, extreme weather phenomena, and sea levels, among other factors, thereby exerting substantial impacts on ecosystems, biodiversity, agriculture, human health, and the economy. Consequently, addressing these critical environmental challenges necessitates immediate scientific attention, urging a reduction in CO_2 emissions across all sectors of human activity (Yoro and Daramola 2020). As the call for reducing global carbon dioxide emissions intensifies, addressing anthropogenic CO_2 emissions becomes increasingly urgent (Kelektsoglou 2018).

To curb such emissions, a legally binding international treaty, the Paris Agreement, was established and came into effect in 2016. Its objective is to constrain average global warming in the twenty-first century to below 2 °C above pre-industrial levels, with a further ambition to strive for limiting warming to 1.5 °C. Over the long term, nations aim to reach the peak of greenhouse gas emissions at the earliest possible juncture, with the ultimate goal of attaining a climate-neutral planet by mid-century (DeConto et al. 2021).

Five years following the ratification of the Paris Climate Agreement, there's been a noticeable slowdown in the growth of global CO_2 emissions. The seismic disruptions brought about by the COVID-19 pandemic have dramatically altered the course of global emissions trajectories. However, the aftermath of COVID-19 has presented a paradox: while there have been shifts towards renewable energy and greener practices, there's also been significant investment in fossil fuel-based infrastructure to aid economic recovery. This dual reality underscores the need for novel policy approaches that can navigate these conflicting interests. As nations strive to recover from the pandemic's fallout, there's a pressing need to ensure that efforts to combat climate change remain steadfast and that post-COVID investments align with long-term sustainability goals (Le Quéré et al. 2021). Balancing economic recovery with environmental stewardship will be paramount in sustaining a downward trend in global emissions beyond the pandemic era.

3.2 GHGs Emissions from Agricultural and Livestock Practices

Numerous greenhouse gases (GHGs) occur naturally in the atmosphere, but human activities exacerbate their accumulation, thereby intensifying global warming. While industrial processes like energy production, petroleum refining, and transportation are major contributors to GHG emissions, a substantial portion stems from highly intensified agricultural and livestock practices (Abdul Latif et al. 2021; Kazancoglu et al. 2021). GHG emissions exhibit a positive correlation with the economic expansion of nations. Additionally, greater support for research and development initiatives from EU countries, coupled with a higher proportion of employment in both industry and agriculture, serves as a catalyst for increased GHG emissions originating from these sectors (Murawska and Goryńska-Goldmann 2023). This underscores the intricate relationship between economic activity, technological advancement, and environmental impact, highlighting the need for comprehensive strategies to decouple economic growth from greenhouse gas emissions.

Livestock farming and intensive agriculture prioritize maximizing production rates, often at the expense of environmental sustainability and animal welfare. Employing practices such as high stocking rates and resource-intensive methods, these activities contribute significantly to climate change. Associated factors like deforestation, inadequate waste management, and excessive fertilizer use further compound their environmental impact. Addressing these issues is imperative for effective climate change mitigation efforts. Urgent attention is needed to promote sustainable practices in these sectors to mitigate their adverse effects on the environment and animal welfare (Tsoraeva et al. 2020).

In recognition of these challenges, the EU Commission has set ambitious targets for GHG emission reductions, requiring each member state to achieve a 30% reduction by 2030 compared to 2005 levels. This directive is part of a broader strategy aimed at gradually transitioning away from super-intensive production models prevalent in the primary sector. By fostering more sustainable agricultural practices and reducing emissions from energy-intensive industries, the EU aims to create a more resilient, competitive, and environmentally responsible energy system. These efforts are not only crucial for meeting short-term climate goals but also lay the groundwork for achieving the EU's long-term vision of substantial GHG reductions by 2050 (EC 2014).

The livestock sector plays a notable role in global GHG emissions, with the meat and dairy industry responsible for a substantial 60% share of emissions attributed to livestock production. Beyond the direct emissions from livestock farming, significant environmental impact arises from land use and land-use changes, particularly deforestation and the conversion of natural habitats for agricultural purposes. Furthermore, emissions associated with food waste further exacerbate the environmental footprint of the meat and dairy sector (Bellarby et al. 2013).

In livestock farming, the emissions with the most significant impact stem from enteric fermentation, accounting for 52.22% of the total emissions, followed by those associated with animal feed, particularly the utilization of imported soybeans. Conversely, emissions related to soil management, manure management, and the use of fossil fuels and electricity are comparatively less significant (Vagnoni and Franca 2018; Escribano et al. 2020). The quantity of purchased feed emerges as a critical factor that negatively influences various environmental outcomes. However, augmenting feed production on farms necessitates enhancements in crop yields and expansion of agricultural land (Zucali et al. 2020).

In assessing the level of intensification in dairy sheep farming GHGs emissions vary significantly, ranging from 1.77 to 4.09 kg CO_2 eq/kg milk. Notably, the lowest emission values are observed in the most intensive farms, where milk production per ewe is maximized. Conversely, the highest emissions are found in more traditional and less productive farming systems. This variability underscores the influence of farming practices and productivity levels on the environmental footprint of dairy sheep production (Batalla et al. 2015; Escribano et al. 2020) (Fig. 3.1).

However, a significant number of studies have omitted soil carbon sequestration from emissions assessments, neglecting the considerable emissions reduction on lower productivity farms attributable to enhanced carbon sequestration from grazing practices (Batalla et al. 2015). Comparisons of carbon footprints per unit of milk from high-yielding dairy systems have revealed that while intensive dairy systems exhibit similar carbon footprints, pasture-based dairy systems demonstrate a lower

Fig. 3.1 Flock of grazing sheep

carbon footprint per unit of milk when factoring in carbon sequestration (O'Brien et al. 2014). Consequently, these carbon sinks play a pivotal role in diminishing net GHG emissions, sequestering between 0.09 and 2.04 kg CO_2 eq/kg milk, thereby substantially attenuating carbon footprint calculations. This underscores the significance of incorporating soil carbon sequestration into emission assessments to accurately evaluate the environmental impact of different farming systems. Moreover, it positions extensive farming systems favorably for potential environmental offsets within the EU policy agenda (Escribano et al. 2020).

3.3 GHGs Emissions from Organic Management Systems

Contemporary conventional agriculture is widely acknowledged as unsustainable and ill-equipped to address critical societal challenges such as climate change, environmental degradation, food security, reliance on fossil fuels, and the erosion of natural resources and biodiversity. In response, there's a notable societal shift towards more sustainable production and management methods, exemplified by extensive and organic farming systems. These approaches prioritize grazing and on-farm food production, resulting in reduced environmental impact compared to intensive farming practices. By predominantly utilizing on-farm resources, extensive systems mitigate energy consumption and lessen dependence on fossil fuels

(EU 848 2018; Rosati et al. 2021). Consequently, in these more extensified systems, the primary sources of GHG emissions often stem from enteric fermentation in ruminant farms or manure management in swine farms, rather than external feed procurement or chemical usage (Buratti et al. 2017).

Key strategies for fostering low-carbon production include reducing energy and water consumption, along with enhancing soil health. Additionally, promoting ecological responsibility across the entire value chain, from production to distribution, consumption, and waste management, is essential (Matthews and Alvez 2012).

However, it's important to consider that while emission calculations may suggest that organic beef systems produce more emissions than conventional ones (Buratti et al. 2017), these assessments often overlook carbon sequestration. In reality, beef and dairy production on pasture, as opposed to intensive grain-fed methods, can lead to reduced GHG emissions (Bellarby et al. 2013) (Fig. 3.2).

Analysis of different systems and species on pasture-based farms, characterized by extensive rearing, has unveiled that cattle farms focusing on fattening calves boast the lowest carbon footprint levels at 8.62 kg carbon dioxide equivalents (CO_2 eq)/kg live weight. They are followed by sheep farms for meat production and those selling calves at weaning. The elevated emissions from enteric fermentation are attributed to the extensification of these systems and the grazing diet of the animals. Notably, feed constitutes the input responsible for the highest percentage of off-farm emissions, accounting for up to 44.60% of total emissions in dairy sheep farms

Fig. 3.2 Extensively grazing beef cattle

and 21.20% in meat sheep farms. In these land-linked systems, akin to organic systems, soil sequestration has been observed to range from 270.02 to 334.01 kg CO_2 eq/ha in extensive farms, serving as a significant carbon offset (Eldesouky et al. 2018).

Table 3.1 depicts the carbon footprint for various production systems and functional units, with the chosen unit expressing the results of the analysis in terms of kg of meat or liter of milk produced (Eldesouky et al. 2018).

In recent research, the correlation between grazing intensity and carbon footprint indicators in dairy sheep farms has emerged, indicating that farms with higher grazing levels tend to exhibit greater environmental sustainability. Notably, these farms

Table 3.1 Carbon footprint of various production systems

Type of product	Production system	Carbon footprint	Functional unit (kg CO_2eq/kg)
Sheep/lamb	Pasture-based	25.9	Lamb live weight
	Mixed system	24.0	Lamb live weight
	Zero-grazing	19.5	Lamb live weight
	Lowland farms	10.85	Lamb live weight
	Upland farms	12.85	Lamb live weight
	Hill farms	17.86	Lamb live weight
	Conventional system	17.5	Lamb meat
	Organic system	10.1	Lamb meat
	Agro-pastoral system—crop-residues	26.6	Carcass weight
	Pastoral system using barley	21.1	Carcass weight
Milk	Semi extensive and local breed	3.8	Corrected milk
	Semi intensive and local breed	3.02	Corrected milk
	Mixed	1.11	Corrected milk
	Intensive and local breed	2.06	Corrected milk
	Extensive and local breed	2.18	Corrected milk
	Extensive system in New Zealand	1	Of energy corrected milk
	Intensive system in Sweden	1.16	Of energy corrected milk
Beef	Extensive system	4.8–8.2	Live weight
	Intensive system	10.6	Live weight
	Extensive (pasture) system	14.0	Live weight
	Conventional system	15.5	Live weight
Pork	Conventional system from feed	6.70–8.70	Live weight
	Conventional system from feed	5.46	Live weight
	Conventional system from feed	3.34	Carcass
Chicken meat	Industrial system	2.6	Carcass weight
	Conventional production system	2.2	Carcass weight
	Free range production system	1.8	Boneless chicken meat portions

demonstrate markedly lower levels of indirect gas emissions and energy consumption compared to their counterparts (Plaza et al. 2021).

These findings underscore the imperative to advocate for livestock models characterized by reduced reliance on external inputs, favoring direct utilization of natural vegetation by animals. To further curtail emissions from such systems, it is essential to optimize the utilization of natural resources while striving for an optimal level of productivity that ensures the economic viability of pastoral livestock farming (Gutiérrez-Peña et al. 2019).

Furthermore, it's worth noting that most agroecological adaptation practices offer synergistic benefits in mitigating GHG emissions and other environmental impacts. Embracing these practices is vital for sustaining food production within the constraints of safe planetary boundaries (Aguilera et al. 2020).

3.4 Carbon Footprint Quantification Through the Life Cycle Assessment

The nexus between climate change and livestock production has underscored the significance of Life Cycle Assessment (LCA)-based carbon footprinting as a global metric for quantifying and reporting GHGs emissions per unit of product (Batalla et al. 2015). LCA serves as a valuable tool for evaluating potential environmental impacts across the entire life cycle of a product, spanning from natural resource acquisition to production, use, and waste management, including disposal and recycling (Finnveden and Potting 2014).

While the carbon footprint offers valuable insights into the contribution of livestock farming to GHG emissions, its application necessitates caution due to methodological complexities. Challenges arise in determining system boundaries, defining the functional unit, and accurately estimating carbon sequestration. Without proper quantification of soil carbon sequestration, there's a risk of overestimating emissions from grazing systems. Therefore, the development of standardized methodologies and models that incorporate soil carbon sequestration in livestock systems is essential. Establishing a specific standardization formula would facilitate accurate carbon footprint calculations and ensure a more holistic assessment of emissions (Gutiérrez-Peña et al. 2019).

3.5 Conclusions

In conclusion, the imperative to curb GHGs emissions, given their role in exacerbating global warming, necessitates concerted efforts across various anthropogenic activities, including livestock farming. Contemporary society is witnessing a notable shift towards more sustainable production and management practices, particularly favoring extensive and/or organic systems. These systems offer a lower environmental footprint compared to intensive farming methods by relying on internal inputs, thereby reducing energy consumption and other resource demands. Moreover, these systems facilitate carbon sequestration in the soil where animals graze, further mitigating emissions.

In light of these considerations, the organic livestock model emerges as a viable strategy for mitigating GHGs emissions. By aligning with the growing global demand for food while upholding environmental sustainability and promoting animal welfare, organic livestock farming represents a holistic approach to addressing the multifaceted challenges posed by climate change and agricultural practices.

References

Abdul Latif SN, Chiong MS, Rajoo S, Takada A, Chun Y-Y, Tahara K, Ikegami Y (2021) The trend and status of energy resources and greenhouse gas emissions in the Malaysia power generation mix. Energies (Basel) 14

Aguilera E, Díaz-Gaona C, García-Laureano R, Reyes-Palomo C, Guzmán GI, Ortolani L, Sánchez-Rodríguez M, Rodríguez-Estévez V (2020) Agroecology for adaptation to climate change and resource depletion in the Mediterranean region. A review. Agric Syst 181:102809

Ahmed N, Khan TI, Augustine A (2018) Climate change and environmental degradation: a serious threat to global security. Eur J Soc Sci Stud

Batalla I, Knudsen MT, Mogensen L, del Hierro Ó, Pinto M, Hermansen JE (2015) Carbon footprint of milk from sheep farming systems in Northern Spain including soil carbon sequestration in grasslands. J Clean Prod 104:121–129

Bellarby J, Tirado R, Leip A, Weiss F, Lesschen JP, Smith P (2013) Livestock greenhouse gas emissions and mitigation potential in Europe. Glob Chang Biol 19:3–18

Buratti C, Fantozzi F, Barbanera M, Lascaro E, Chiorri M, Cecchini L (2017) Carbon footprint of conventional and organic beef production systems: an Italian case study. Sci Total Environ 576:129–137

DeConto RM, Pollard D, Alley RB, Velicogna I, Gasson E, Gomez N, Sadai S, Condron A, Gilford DM, Ashe EL, Kopp RE, Li D, Dutton A (2021) The Paris climate agreement and future sea-level rise from Antarctica. Nature 593:83–89

EC (2014) 2030 Climate ad eergy policy framework. European Council 169(14):1–16

Eldesouky A, Mesias FJ, Elghannam A, Escribano M (2018) Can extensification compensate livestock greenhouse gas emissions? A study of the carbon footprint in Spanish agroforestry systems. J Clean Prod 200:28–38

Escribano M, Elghannam A, Mesias FJ (2020) Dairy sheep farms in semi-arid rangelands: a carbon footprint dilemma between intensification and land-based grazing. Land Use Policy 95:104600

EU 848 (2018) Regulation (EU) 2018/848 of the European Parliament and of the Council of 30 May 2018 on organic production and labelling of organic products and repealing Council Regulation (EC) No 834/2007

Finnveden G, Potting J (2014) Life cycle assessment. In: Encyclopedia of toxicology. Elsevier, pp 74–77

Gutiérrez-Peña R, Mena Y, Batalla I, Mancilla-Leytón JM (2019) Carbon footprint of dairy goat production systems: a comparison of three contrasting grazing levels in the Sierra de Grazalema Natural Park (Southern Spain). J Environ Manag 232:993–998

Kazancoglu Y, Ozbiltekin-Pala M, Ozkan-Ozen YD (2021) Prediction and evaluation of greenhouse gas emissions for sustainable road transport within Europe. Sustain Cities Soc 70:102924

Kelektsoglou K (2018) Carbon capture and storage: a review of mineral storage of CO_2 in Greece. Sustain For 10:4400

Le Quéré C, Peters GP, Friedlingstein P, Andrew RM, Canadell JG, Davis SJ, Jackson RB, Jones MW (2021) Fossil CO_2 emissions in the post-COVID-19 era. Nat Clim Chang 11:197–199

Matthews AG, Alvez J (2012) Sustainability indicators for low carbon farming. In: Rahmann G, Godinho D (eds) Tackling the future challenges of organic animal husbandry. vTI Agriculture and Forestry Research, Hamburg

Murawska A, Goryńska-Goldmann E (2023) Greenhouse gas emissions in the agricultural and industrial sectors—change trends, economic conditions, and country classification: evidence from the European Union. Agriculture 13:1354

O'Brien D, Capper JL, Garnsworthy PC, Grainger C, Shalloo L (2014) A case study of the carbon footprint of milk from high-performing confinement and grass-based dairy farms. J Dairy Sci 97:1835–1851

Plaza J, Revilla I, Nieto J, Hidalgo C, Sánchez-García M, Palacios C (2021) Milk quality and carbon footprint indicators of dairy sheep farms depend on grazing level and identify the different management systems. Animals 11

Rosati A, Borek R, Canali S (2021) Agroforestry and organic agriculture. Agrofor Syst 95:805–821

Tsoraeva E, Bekmurzov A, Kozyrev S, Khoziev A, Kozyrev A (2020) Environmental issues of agriculture as a consequence of the intensification of the development of agricultural industry. E3S Web Conf 215:02003

Vagnoni E, Franca A (2018) Transition among different production systems in a Sardinian dairy sheep farm: environmental implications. Small Rumin Res 159:62–68

VijayaVenkataRaman S, Iniyan S, Goic R (2012) A review of climate change, mitigation and adaptation. Renew Sust Energ Rev 16:878–897

Yoro KO, Daramola MO (2020) CO_2 emission sources, greenhouse gases, and the global warming effect. In: Advances in carbon capture. Elsevier, pp 3–28

Zucali M, Lovarelli D, Celozzi S, Bacenetti J, Sandrucci A, Bava L (2020) Management options to reduce the environmental impact of dairy goat milk production. Livest Sci 231:103888

Chapter 4
Indigenous Poultry in Indonesia

Wisje Lusia Toar, Laurentius J. M. Rumokoy, Carlos Palacios Riocerezo, and Manuel Gonzalez Ronquillo

Abstract This chapter explores the production of indigenous Indonesian poultry, with a particular focus on chickens. It aims to offer an overview of the diversity of Indonesia's native poultry breeds and their production practices. The content of this chapter draws from direct field experience and is enriched by practical knowledge as well as scientific and technical references from both online and print sources. It provides a general perspective on Indonesia's indigenous chickens, covering both wild and domesticated varieties and highlighting their unique characteristics. The following section then examines the care practices associated with free-range and caged rearing methods relevant to poultry production.

Keywords Native chicken · Organic · Nature · Indonesia · Small scale business

W. L. Toar (✉)
Department of Nutrition and Feed Technology, Faculty of Animal Science, Sam Ratulangi University, Manado, Indonesia

L. J. M. Rumokoy
Department of Animal Production, Faculty of Animal Science, Sam Ratulangi University, Manado, Indonesia
e-mail: rumokoy@msn.com

C. Palacios Riocerezo
Area of Animal Production, Faculty of Agricultural and Environmental Sciences, University of Salamanca, Salamanca, Spain
e-mail: carlospalacios@usal.es

M. Gonzalez Ronquillo
Facultad de Medicina Veterinaria y Zootecnia, Universidad Autonoma del Estado de Mexico, Toluca, Mexico
e-mail: mrg@uaemex.mx

C. Palacios Riocerezo, M. Gonzalez Ronquillo (eds.), *Traditional Livestock Production*, https://doi.org/10.1007/978-3-031-88275-3_4

4.1 General Points of View

The production of indigenous Indonesian poultry is influenced by various factors, including human and natural resources, along with climatic conditions that impact the availability of plant and animal nutrients. These nutrient sources are consumed directly by the chickens and also supplemented by breeders. Economically, the outlook for Indonesia‘s indigenous poultry sector is promising, as strong consumer demand for meat and eggs offers significant growth potential. However, challenges persist, such as financial constraints, geographical limitations, and Indonesia’s distinct climate. Despite these obstacles, indigenous poultry species are sustained and preserved by local farming communities.

4.2 Biodiversity

Indonesia‘s biodiversity varies significantly across its many regions, even within individual islands. The viability of poultry farming in this archipelagic nation depends on numerous factors, such as climate, rainfall, temperature ranges, and geographic location. Indonesia comprises over 17,000 islands, spanning three time zones. Its five largest islands—Kalimantan, Sumatra, Papua, Sulawesi, and Java—each feature unique ecological landscapes. A diverse array of wild bird species can be found throughout these islands, although certain species are endemic to specific locations.

Indonesia’s indigenous poultry, both wild and domesticated, represent a valuable natural resource, including chickens, ducks, and various bird species. Some of these animals, such as certain types of jungle fowl, are classified as endemic and are legally protected. Notably, native Indonesian chickens like the green jungle fowl exhibit distinctive appearances, as shown in Fig. 4.1.

Fig. 4.1 A type of Indonesian forest chicken native to the region (Amiru 2023). Two images of a colourful rooster with a vibrant red comb and iridescent feathers. The left image shows the rooster perched on a branch surrounded by green leaves. The right image depicts the rooster standing on lush green grass

Fig. 4.2 A vibrant rooster with multi-coloured plumage stands on a tree branch in a forest setting, displaying its red comb and wattles. To the right, a hen with muted brown and grey feathers is standing on the ground amidst green foliage. ("The-Animal-Fact" retrieved in December 2024)

Fig. 4.3 Bali duck or penguin duck (Roys-Farm 2023). Four white ducks with fluffy crests on their heads standing in a row, facing right. They are in a grassy area enclosed by a wire fence. The background is filled with greenery

Figure 4.2 represents a red jungle fowl, a bird species found on various Indonesian islands, is the direct ancestor of the domesticated chicken and also inhabits regions across Asia (Cakrawala 2021). In addition to chickens, Indonesia is known for a variety of notable duck breeds, such as the Tegal, Mojosari, Bali (in Fig. 4.3) and Alabio.

4.3 Indonesian Indigenous Poultry

4.3.1 Bali Duck or Pinguin Duck (Anas Sp.)

Bali ducks, also known as Balinese ducks, are a species of waterfowl native to the islands of Bali and Lombok.

Despite being classified as endangered, these birds are notable for their high disease immunity. As shown in the accompanying image, the male's tail has a distinctive curve. These ducks, often raised intensively, produce about 250 eggs per year. Each egg weighs approximately 70 grams on average, which is smaller than the Alabio duck's eggs (Podomoro-Feedmill 2021).

4.3.2 *Alabio Duck*

The duck in question is known as the native Indonesian duck. (Hetzel 1984; Sulaiman et al. 2022). The name "Alabio" (in Fig. 4.4) is derived from the name of the district where these ducks originated. However, they have been crossed with Pekin ducks. (Howell 2021).

The Alabio duck is renowned for its high-quality eggs and tender meat. This Indonesian breed, native to the southern region of Kalimantan, can produce up to 250 eggs annually when raised in a cage-breeding system. The egg production period for these ducks can last up to three years (Agrozine 2023).

Fig. 4.4 A group of ducks with brown and white feathers and orange beaks. The top image displays a large flock of ducks closely packed together, while the bottom image focuses on a smaller group of ducks peering through wooden slats. The setting appears to be a farm or enclosure. (Howell 2021)

Fig. 4.5 *Macrocephalon maleo*. Two black and orange birds with blue and yellow markings on their heads on a sandy ground. The background shows a natural setting with soil and scattered leaves. (Source: *Ilmu-Pengetahuan.org 2014.* Retrieved in January 2023)

Fig. 4.6 *Macrocephalon maleo* (Wicaksono 2024). A pair of Macrocephalon maleo, male and female, walks across the sand. This species is known for its lifelong loyalty to a single mate

4.3.3 Burung Maleo (Macrocephalon maleo)

In this section, we will introduce another species of wild bird, the maleo (*Macrocephalon maleo*), also known as 'burung maleo'. The image of this species is depicted in Figs. 4.5 and 4.6. This species is native to the island of Sulawesi, where its populations are found in several regions, including North Sulawesi, Gorontalo, Central Sulawesi, Southeast Sulawesi, West Sulawesi, and South Sulawesi. The maleo is considered an endemic species and has been classified as a protected animal by the government in accordance with relevant regulations. The bird in question, identified by UU RI Number 5, was observed in 1990. Additionally, Government Regulation Number 7 of 1999 is worth noting. During the regeneration process, the bird was observed laying its eggs in sand or on the ground. After hatching, the young bird was seen attempting to escape from the nest and move to a safer area. The maleo bird's egg weighs approximately 210 grams, with an average length of 10 cm and width of 6 cm (Widnyanya 2020). The bird's habitat has been significantly reduced due to the conversion of its natural environment into plantations and resettlement areas.

According to Saerang et al. (2011), the Macrocephalon maleo is a monogamous species, with pairs—consisting of a male and a female—remaining lifelong partners. This bird is typically found in forested habitats, though it has also been observed in agricultural areas. Due to its endangered status, it is illegal to capture or

Fig. 4.7 Maleo bird chicks in a cultivation cage (Beritapalu 2023). Five young birds are inside a blue wire cage. They are gathered around a yellow feeder filled with grains. A white water dispenser with red base is on the right side. The setting appears to be a controlled environment, likely for raising or studying the birds

Fig. 4.8. A vibrant bird of paradise with striking yellow and green plumage and long, flowing tail feathers perched on a branch. The background is a blurred mix of green foliage, suggesting a tropical environment. (Hewan-Kartun 2021)

kill this species, with habitat loss from deforestation being a primary contributing factor. Beritapalu (2023) reported that the species' conservation can be supported through population increases, which may be achieved through breeding programs in "captive systems."

Figure 4.7 shows a young maleo observed in a specialized breeding cage. The observer, Mobius Tanari, successfully bred maleo birds by raising dozens of them in a dedicated cage located at the side of his residence (Beritapalu 2023). The eggs of this bird can be hatched using hatching machine technology (Saerang et al. 2011).

4.3.4 Bird-of-Paradise (Cendrawasih)

The family of Paradisae, which is also known as the bird-of-paradise (*Cendrawasih*), comprises 15 species (Gill et al. 2022) that are endemic to Papua Island, occurring in both Papua Indonesia and Papua New Guinea. A species of this family is depicted in Fig. 4.8. This bird has become a symbol of greatness for various tribes in the Papua region.

The population of this bird continues to decline, highlighting the urgent need for effective conservation strategies. Cita (2016) suggests that the Paradiseae bird family's population can be increased through in situ conservation and management practices, which include providing care in controlled environments such as cages. Paradisaea apoda, a species of the Paradisaeidae family, is commonly found in Indonesia. Other species from this family have been observed on the islands of Maluku, Halmahera, and in eastern Australia.

4.3.5 Domesticated Indigenous Chickens

A variety of indigenous Indonesian chicken breeds have been domesticated and selectively bred by local communities. Raising chickens is a common practice in rural and suburban areas, often serving as a supplementary activity within household economies. Some of these indigenous chickens are named based on their geographic origins and distinctive colorations. Indonesia is home to numerous native chicken species, several of which are particularly well-known. These include the Minahasa Jungle Chicken, Sumatran Chicken, Cemani Chicken, Gray Jungle Chicken, Green Jungle Fowl, Bekisar Chicken, Red Jungle Fowl, Gaok Chicken, and Pelung Chicken. Figure 4.9 shows indigenous and hybrid indigenous chickens cultivated in traditional farming in North Minahasa Regency.

The number of households that raise indigenous chickens is relatively high, though the number of adult chickens kept is typically limited to fewer than 50. These chickens are raised using simple methods, with the farming practices passed down through generations. This agricultural approach is known as 'traditional farming'.

Fig. 4.9 The Indonesian indigenous chickens. Two images of chickens with an orange arrow pointing from left to right. The left image shows an "indigenous" chicken standing on concrete. The right image shows a "hybrid indigenous" chicken on grass

Fig. 4.10 Indonesian Tukang chickens, Tribudi et al. (2023). The left image shows a rooster with white and black feathers and a hen with black feathers on a dirt ground. The hen has an orange band on its leg. The right image shows a young chicken with mottled feathers standing against a dark background. Tribudi et al. (2023)

The number of households that keep indigenous chickens is relatively high, but the number of adult chickens kept is limited to less than 50. These chickens are bred in a straightforward manner, with the maintenance pattern being carried over from generation to generation. This agricultural practice is referred to as "traditional farming". A day-old chick (DOC) is the result of natural brooding. In general, indigenous chickens that are kept in an "open system" can traverse the boundaries of their owner's yard, thereby facilitating interaction and mating with other chickens that have a history of mixed ancestry.

This condition has been in existence for an extended period of time, during which time the indigenous chickens have been hybridized with other breeds. One of the defining characteristics of Indonesian indigenous chickens is the diversity of their feather patterns, which are not uniform. In order to maintain the genetic authenticity of indigenous chickens in Indonesia, a conservation pattern is employed in accordance with the recommendations set forth by Tribudi et al. (2023) on the Tukong chicken (Fig. 4.10), which originates from the island of Kalimantan.

Various efforts have been made by Indonesian native chicken breeders to increase production, including crossbreeding between native Indonesian chickens and superior commercial breeds. The results of this crossbreeding have started to attract attention from both the farming community and academics. Figure 4.11 shows a photo of a crossbreeding result. This photo was taken at "Sentrum Agraris Lotta" (SAL) in Pineleng.

4.4 Benefit of Indigenous Chickens Farm

The financial investment required to raise indigenous chickens is significantly less than that required to raise purebred chickens. This is due to a number of benefits that are associated with the farmer, including the following:

Fig. 4.11 Native Indonesian chickens that have been crossed with superior hybrid chickens. Four chickens standing on a tiled floor in front of a metal door. The chickens are of various colors, including brown, black, and white. They are closely grouped together, facing different directions

4.4.1 Entertainment

Some indigenous poultry farmers engage in chicken-raising as a hobby or for their own personal enjoyment, with the objective of deriving pleasure from the activity itself, rather than seeking financial profit. This type of breeder is primarily concerned with the well-being of his animals and is willing to invest a significant amount of his personal resources in order to ensure their optimal health and appearance. Conversely, some poultry breeders maintain these animals as a hobby and as a means of generating financial profit from their production.

Conversely, indigenous chickens are also kept as an object of entertainment through contest activities as shown in Fig. 4.12. The objective of the show is multifaceted, encompassing the demonstration of the distinctive crowing sounds of a rooster, as well as the showcasing of the unique color and physical characteristics of the chicken.

The crowing of a rooster was accompanied by the sound of laughter, which could be described as a prolonged vocalization akin to singing. The variety of sounds produced by these indigenous chickens is on display at certain contest events. Conversely, there are those who maintain indigenous chickens for engaging in combat activities, as exemplified in the accompanying illustration (Fig. 4.13). As long as it is solely for exhibition activities and has the necessary permits from the authorities for the exhibition, this activity can be carried out.

However, the legal framework does not permit activities involving gambling. Consequently, if a community member engages in cockfighting in conjunction with gambling, they are in violation of the regulations and will be held legally responsible.

Fig. 4.12 Three roosters perched on wooden stands in front of a building. The central rooster is black and white with a prominent red comb, while the others have red and black plumage. The background shows a structure with windows and a roof. (Source: https://www.youtube.com/watch?v=Oe9odIzw4QY)

Fig. 4.13 Two roosters engaged in a mid-air fight, with wings spread and claws extended. The rooster on the left is predominantly black, while the one on the right has a mix of red and brown feathers. The background is a blurred natural setting with green foliage. (Source: Beritatangsel 2023)

4.4.2 Meat and Egg Producer

The advantage of free-range chicken production conducted by local farmers in Indonesia is that it contributes to the community food supply. Eggs and meat produced from these chickens are usually limited to family consumption. If this production exceeds the needs of the farmer's family, it can be sold to local markets. This can be traditional markets, supermarkets, or food depots around the farm. The production is relatively small quantity, so there is not enough to be on sale nationally to other provinces.

4.4.2.1 Culinary Benefits

Indigenous chicken has a distinctive taste and is more popular with consumers than purebred chicken. This flavor is related to several factors, including feed and space: Unlike the broiler and layer industries, native chickens do not rely on commercial feed. Farmers provide only a minimal amount of feed made from kitchen waste or sourced from non-continuous agricultural products, such as the supply of corn or rice bran local chickens like to hunt for their own food from the surrounding natural environment, such as insects and other species of animals for feed that is lower in fat than commercial feed. Outdoor exercise allows these chickens to use their energy for running, foraging or for the social activities of the chickens themselves, resulting in limited accumulation of fat in the adipose tissue. This fact has a positive impact on the taste of the local chicken dishes that are typically found in the different regions of Indonesia.

4.4.2.2 Economic Benefits

One of the attractions for tourists carrying out ecotourism activities in Indonesia is to explore the unique of biodiversity found in areas spread across Indonesia, including various species of wild birds that have been conserved in zoos or other places that have authorization to carry out conservation that can be seen by visitors. Foreign tourists also help to increase the country's foreign exchange. The support of domestic tourists could contribute to the improvement of the income level of the local community.

4.4.3 The Rearing Patterns of Indigenous Indonesian Chickens

In general, the care of indigenous chickens in Indonesia is carried out by farmers or as a side activity to other main activities of a household whose care is carried out traditionally, that is, simply following the habits of the ancestors. This pattern of raising chickens is found in rural areas to suburban areas, even sometimes you can find chickens that are cultivated by farmers in their domicile locations: ornamental chickens, or fighting chickens. The maintenance technique is varied, which is not bound to the practical and scientific guidelines. Some of the maintenance systems used are as follows:

4.4.3.1 Closed Maintenance System

During the starter period, indigenous chickens are mainly kept in a closed system. The chickens are kept in closed crates to prevent them from wandering, and this system allows the chicks to be separated from their mother to reproduce more

Fig. 4.14 (**a**) A hen sitting on a nest inside a wooden coop. (**b**) Several chicks under a warm light bulb in a small enclosure. (**c**) A larger wooden brooder with many chicks, equipped with multiple light bulbs and feeders

Fig. 4.15 Image showing two scenes, (**a**) and (**b**): The left, labeled **a**, shows several chickens gathered inside a wooden coop. The right, labeled **b**, shows a person standing in front of pink cages. The setting suggests a farm or poultry environment

quickly. Chickens in cages are dependent only on the food supply provided by the breeder. This maintenance pattern is depicted in Figs. 4.14 and 4.15. Their growth and development are determined to a large extent by the attention and skill of the farmer with regard to the environment, nutrient amount.

Fig. 4.16 Two images labeled **a** and **b**. Image **a** shows two chickens pecking at the ground on a wet, muddy surface. The chicken on the left has vibrant red and black feathers, while the one on the right is mostly black. Image **b** depicts a hen with brown feathers on a grassy area, surrounded by three small chicks, indicated by orange arrows. The ground is lush and green

The main feeds for the chickens are maize grains, rice bran, agricultural products or kitchen waste. In indigenous intensive or semi-intensive chicken farming, some farmers use simple incubators to hatch the eggs. For this reason, it is also necessary for the farmer to prepare cage units intended for adult chickens.

4.4.3.2 Opened Maintenance System

In most small-scale indigenous chicken farming, an open management system is found. Insects and various natural resources available in the open environment, including various types of insects (Rumokoy et al. 2022; and Toar et al. 2018), are a food source for chickens that are active in the open system. Chickens move freely for all their activities: eating what is found around the farm, foraging for food on the ground, under plants and hunting for prey as figured in Fig. 4.16.

Fig. 4.17 Three images labeled **a**, **b**, and **c**. Image **a**: Overgrown green vegetation with two yellow circles highlighting a rooster on the left and a small structure on the right. Image **b**: Closeup of a grasshopper on a green leaf. Image **c**: Closeup of green chili peppers on a plant with a small insect on a leaf

This activity positively influences muscle development for meat production and reduces adipose fat accumulation, as energy is expended during continuous outdoor activity. Consequently, this condition promotes leaner chicken products, aligning well with consumer preferences for taste.

Insects can serve as a source of nutrition for free-ranging birds (Toar et al. 2022), as illustrated in Fig. 4.17. However, this type of chicken has several weaknesses, including poor health management, which increases their risk of exposure to pathogens such as harmful microbes and various parasites that can cause illness or even death, leading to losses for farmers. Indigenous chickens, however, are generally better adapted to these environmental conditions than purebred chickens. This resilience may be related to genetics and environmental food sources, including immune-boosting substances found in insects (Toar et al. 2019). On the other hand, free-ranging chickens can easily wander beyond their owner's property and potentially damage neighbouring plants. To prevent this, owners need to control their chickens to keep them within their designated area.

Acknowledgments This scientific paper (Book Chapter) is an outcome of the international publication and research collaboration between Sam Ratulangi University, Indonesia, and the University of Salamanca, Spain, specifically realized at the USAL Faculty of Environmental and Agricultural Sciences. This work was funded by the Institute for Research and Community Service (LPPM) of Sam Ratulangi University under the Unsrat RKLN - PNBP-BLU Project for the academic year 2024.

References

Agrozine (2023) Mengenal Itik Alabio dan Tips Budidayanya. https://agrozine.id/mengenal-itik-alabio-dan-tips-budidayanya. Retrieved 28 January 2023

Amiru N (2023) 7 Fakta Unik Ayam Hutan Hijau, Pandai Terbang dan Punya Suara Khas. https://bobo.grid.id/read/083671110/7-fakta-unik-ayam-hutan-hijau-pandai-terbang-dan-punya-suara-khas?page=all. Retrieved 01 February 2023

Beritapalu (2023) 'Melihat Burung Maleo Tak Harus ke Banggai, di Palu pun Bisa'. https://beritapalu.com/2023/08/01/melihat-burung-maleo-tak-harus-ke-banggai-di-palu-pun-bisa. Retrieved 26 Januari 2023

Beritatangsel (2023) https://beritatangsel.com/2023/12/22/ketum-fwj-indonesia-sayangkan-judisabung-ayam-di-malang-tak-tersentuh-aparat/. Retrieved 03 February 2023

Cakrawala (2021) Ayam hutan merah nenek moyang ayam peliharaan. https://dody94.wordpress.com/2011/04/01/ayam-hutan-merah-nenek-moyang-ayam-peliharaan/2. Retrieved 26 Januari 2023

Cita KD (2016) Desain penangkaran cendrawasih kecil (*Paradisaea minor* Shaw, 1809). Jurnal Nusa Sylva 16(2):65–71

Gill F, Donsker D, Rasmussen P (eds) (2022, January) Crows, mudnesters, melampittas, Ifrit, birds-of-paradise. IOC World Bird List. https://www.worldbirdnames.org/bow/crows/

Hetzel DJ (1984) Comparative performance of intensively managed Khaki Campbell and native Indonesian ducks. Trop Anim Health Prod 16(1):39–46

Hewan-Kartun (2021) Istimewa Burung Cendrawasih. https://hewankartun.blogspot.com/2021/11/istimewa-burung-cenderawasih-gambar.html. Retrieved 28 January 2023

Howell C (2021) Alabio Duck Breed: everything you need to know about this birds. https://www.thehappychickencoop.com/alabio-duck-breed-everything-you-need-to-know-about-this-birds/. Retrieved 03 February 2023

IlmuPengetahuan.org (2014) Burung Maleo, Hewan Endemik Pulau Sulawesi. https://ilmupengetahuan.org/burung-maleo-hewan-endemik-pulau-sulawesi/. Retrieved 01 February 2023

Podomoro-Feedmill (2021) Itik Bali Jambul yang Terancam Punah. https://podomorofeedmill.com/info/itik-bali-jambul-yang-terancam-punah. Retrieved 28 January 2023

Roys-Farm (2023) Bali Duck characteristics, origin & uses info. https://www.roysfarm.com/bali-duck. Retrieved 03 February 2023

Rumokoy L, Pudjihastuti E, Rumokoy D, Kiroh H, Ngangi LR, Rawung V, Toar W (2022) Growth performance of 'super native chicken' treated with a suplementation of maggot flour of *Drosophila melanogaster* in ration. https://animalsciencejournal.usamv.ro/pdf/2022/issue_2/Art20.pdf

Saerang JL, Rawung VR, Lucia L (2011) Teknologi Penetasan Burung Maleo (*Macrocephalon maleo*) Sebagai Upaya untuk Mengatasi Kepunahan. In: Seminar Nasional Petemakan Berkelanjutan Ke-3 Fakultas Petemakan Universitas Padjadjaran "Road To Green Farming". Universitas Padjadjaran-Jatinangor, pp 111–114

Sulaiman A, Rahmatullah S, Effendi H, Simanungkalit G (2022) Production performance of Alabio ducks (Anas platyrhynchos Borneo) under different levels of drinking water salinity. J Adv Vet Anim Res 9(2):241–245

The-Animal-Fact. Red Junglefowl Fact File. https://www.theanimalfacts.com/birds/red-junglefowl. Retrieved 23 December 2024

Toar WL, Tulung M, Memah V, Pudjihastuti E, Rumokoy L, Untu IM (2018) The presence of insects in animal farm in North Sulawesi. Sci Papers Ser D Anim Sci 61. https://animalscience-journal.usamv.ro/pdf/2018/issue_1/Art39.pdf

Toar WL, Rumokoy LJM, Pudjihastuti E, Manangkot H, Bagau B, Untu IM (2019, November). Effect of supplementation of combination of curcuma and BSF maggot meal in rations on accumulative weight of native chickens. In: IOP conference series: earth and environmental science, Vol. 372, No. 1. IOP Publishing. p 012009. https://doi.org/10.1088/1755-1315/372/1/012009

Toar W, Pudjihastuti E, Turangan S, Assa G, Sompie F, Rumokoy L (2022) The effect of supplementation of Patanga succincta flour in ration on indigenous chickens meat production. https://animalsciencejournal.usamv.ro/pdf/2022/issue_2/Art21.pdf

Tribudi YA, Natsir MH, Ulfah M, Gusmalawati D, Sari APZNL, Hariyono DNH, Prihandini PW, Nurgiartiningsih VMA (2023) Tukong: a rumpless indigenous chicken breed from West Kalimantan, Indonesia. Worlds Poult Sci J 79:1–15

Wicaksono RA (2024) Maleo Tak Pernah Ingkar Janji. In. Konservasi-Biodiversitas (Editor Yosep Suprayogi). https://papua.betahita.id/news/detail/9994/maleo-tak-pernah-ingkar-janji.html.html

Widnyanya IGNP (2020) Kajian morfologi dan morfometri telur burung Maleo (*Macrocephalon maleo* SAL. Muller 1846). Agropet 14(1):31–37

Chapter 5
Insect Use in Organic Animal Production Systems

Jaime Nieto de la Losa, Javier Plaza Martín, and Carlos Palacios Riocerezo

Abstract Meat consumption is progressively increasing to meet the dietary needs of a growing world population with increasing purchasing power, thus altering their dietary patterns. In addition, there is also a growing concern for environmental sustainability and animal welfare linked to food production. In this line, there is a widening sector of the population opposing certain conventional raw materials used in the feed industry, due to their lack of environmental sustainability and high production costs. Therefore, it is becoming necessary to look for new sources of protein to feed not only humans, but also animals.

Particularly, soybean meal is the most widely used protein source in animal feed, both in ruminants and monogastrics, but the increase in its price together with the environmental problems related to its production has led to seek new alternatives to replace it in livestock production. Its cultivation causes serious deforestation problems worldwide, enormous water consumption, indiscriminate use of pesticides and chemical synthesis products, as well as an irrecoverable loss of biodiversity. Furthermore, most of its production takes place in the Americas, where genetically modified varieties are used, thereby making its use prohibitive in organic production systems, where it is even more important to look for alternative protein sources to avoid serious social, economic and ecological impacts.

Among the alternative protein sources used in animal feed, insect meal is considered as a potential substitute for soybean meal, since it has a significant amount of unsaturated fats and high concentrations of vitamins, fiber and minerals. As an added value, its environmentally sustainable production stands out, given its low soil and water requirements, low carbon footprint and its relation with circular economy models. In addition, its production rate is high due to the short life cycle of insects and its production system is relatively simple.

The EU recently authorized the use of processed animal protein derived from farmed insects in aquaculture production. This feed source, intended for the

J. Nieto de la Losa (✉) · J. Plaza Martín · C. Palacios Riocerezo
Area of Animal Production, Faculty of Agricultural and Environmental Sciences, University of Salamanca, Salamanca, Spain
e-mail: jaimenl@usal.es; pmjavier@usal.es; carlospalacios@usal.es

C. Palacios Riocerezo, M. Gonzalez Ronquillo (eds.), *Traditional Livestock Production*, https://doi.org/10.1007/978-3-031-88275-3_5

production of feed for farm animals different from fur animals, must meet a number of safety constraints that restrict the usable species to the following: black soldier fly (*Hermetia illucens*), common house fly (*Musca domestica*), mealworm (*Tenebrio molitor*), bed beetle (*Alphitobius diaperinus*), house cricket (*Acheta domesticus*), striped cricket (*Gryllodes sigillatus*) and bicolor cricket (*Gryllus assimilis*). In addition, insects are gaining commercial interest since the even more recent EU regulation of 2021, which allowed the use of insect meal in diets for poultry, pigs and fur animals, as well as aquaculture animals.

Organic agriculture already uses insects as a biological control method to fight against different pests, which, added to their nutritional characteristics and the sustainable production, suggests a promising future for insects to be used in the rations of animals raised under the organic production regime.

Keywords Alternative protein sources · Insect meal · Insect inclusion · Organic feed · Sustainability

5.1 Introduction

Global meat consumption is on the rise, primarily fueled by economic growth and shifting dietary patterns, especially in developing nations. This economic expansion results in higher incomes, enabling more individuals to enjoy a diverse and affluent diet that includes meat. Additionally, the emergence of a burgeoning middle class and urbanization often correlates with a preference for Western diets, which typically feature a higher meat intake compared to traditional dietary practices. The pursuit of balanced and healthier eating habits further underscores the importance of incorporating meat as a rich source of high-quality protein. Moreover, advancements in agricultural and livestock technologies, coupled with improvements in the supply chain, have significantly enhanced the availability and accessibility of meat products, further driving up consumption levels (Allegretti et al. 2018).

Indeed, the surge in meat consumption gives rise to apprehensions regarding its implications for both human health and the environment. Issues such as deforestation, greenhouse gas emissions, and biodiversity loss are increasingly brought to the forefront. Additionally, ethical considerations come into play, particularly concerning the welfare of animals, particularly those subjected to intensive farming practices (Thornton 2010).

The rise in meat production, driven by heightened consumption, fosters the intensification of livestock farming practices. This shift entails reduced reliance on natural environments and an increased dependence on imported energy and raw materials. Consequently, there's a decline in extensive livestock farming methods, characterized by animals being raised across larger areas and heavily reliant on local resources (Ruiz Mirazo 2011). This transformation occurs concurrently with the dwindling availability of natural resources and rural populations, alongside escalating demands for sustainable agricultural practices. Addressing these external challenges and factors impacting production efficiency necessitates the development of innovative solutions (Hovhannisyan and Grigoryan 2016).

Addressing these challenges requires the promotion of responsible and ethical farming practices within meat production. Organic livestock farming stands as a champion of these principles, prioritizing environmentally sustainable production, the preservation of animal health, and adherence to elevated welfare standards, all while upholding the quality of the products obtained. By steadfastly pursuing these aims, organic livestock farming effectively caters to the growing demands of an increasingly discerning consumer base that scrutinizes conventional production methods (Sundrum 2001).

Furthermore, escalating resource scarcity for feed production and mounting environmental apprehensions underscore the unsustainable nature of conventional feed sources. Hence, there arises a pressing need to explore more sustainable alternatives, encompassing practices that enhance soil health and diminish dependence on external inputs (de Souza et al. 2021).

Effectively tackling the unsustainability of conventional feed sources demands a comprehensive approach that advocates for natural resource conservation, environmental preservation, and the welfare of animals (Abro et al. 2020).

5.2 Raw Materials Employed in Organic Farming

In all livestock farming operations, feeding plays a pivotal role, influencing both production outcomes and animal welfare. The feed utilized in organic livestock farming is governed by Article 8 of Regulation (EU) 2018/848, which delineates specific principles applicable to the processing of organic feed. According to these regulations, processed organic feed must be derived from organic feed materials, with strict limitations imposed on the use of additives and processing aids. These additives and aids should only be employed to the minimum extent necessary, solely for essential technological or zootechnical purposes, or for specific nutritional needs. Moreover, substances and processing methods that could potentially misrepresent the true nature of the product are expressly prohibited. Furthermore, the processing of feed must adhere meticulously to organic standards, employing predominantly biological, mechanical, and physical methods (EU 848 2018).

In recent decades, the growing consumer interest in organically produced foods has spurred the necessity to align global livestock production with preferences for more natural and sustainable products (Kushwah et al. 2019) (Fig. 5.1).

However, a notable challenge arises from the scarcity of permitted feed ingredients and the limited range of feed additives available on the market due to stringent organic regulations. This dearth of options, compounded by the challenges faced by the feed industry in transitioning to organic practices, hampers the advancement of sustainability within the livestock sector. This impact is felt not only by livestock companies and feed mills already operating under organic standards but also by those still entrenched in conventional production systems (Escribano 2018).

Fig. 5.1 Slow-growing, free-range chickens

The growing scarcity of resources for feed production is compounded by escalating environmental apprehensions surrounding its cultivation. There's a rising concern over raw material competition between food and feed, as well as challenges related to input availability. Feed production necessitates cropland, along with fertilizers, water, and energy (Flachowsky et al. 2017; Abro et al. 2020; te Pas et al. 2021), resulting in heightened costs.

5.3 Protein Sources Problematic

Protein sources utilized in livestock feed represent one of the most expensive and constraining feed ingredients (Selaledi et al. 2020). Hey constitute a vital component in the diets of all animal species, playing a pivotal role in their development, growth, and reproductive efficiency, thereby influencing their overall health, welfare, and productive performance (Kim et al. 2019). Soybean meal stands out as the primary protein source in animal rations, commanding a significant share of the global market, although other legumes such as peas, lentils, and chickpeas also serve as important sources of dietary protein (Boye et al. 2010).

The bulk of soybean meal production is concentrated in the Americas, resulting in a high level of dependency on soybean imports within the European Union's livestock sector. This reliance on imported soybean gives rise to challenges such as intermittent supply, limited availability, elevated costs, trade disputes, and geopolitical factors, underscoring the imperative to diversify protein sources (Khan et al. 2023).

Furthermore, the importation of soybean is frequently associated with issues like deforestation and genetic modification, among others, which are subject to significant scrutiny by consumers (Soisontes et al. 2023). Hence, it becomes imperative to optimize livestock production to mitigate dependency on imported soybean feed, thereby reducing the demand for cropland in deforestation-prone regions while satisfying the nutritional requirements of EU diets (Karlsson et al. 2020).

The rapid expansion of soybean cultivation as the primary protein source for animal feed has sparked concerns within society (Vauterin et al. 2021), giving rise to significant environmental issues (Sánchez Muros et al. 2014).

The cultivation of soybeans contributes to deforestation in ecologically valuable areas, threatening forests and savannas (Fehlenberg et al. 2017) and jeopardizing the biodiversity of these regions (Richards et al. 2012). Moreover, soybean cultivation demands substantial quantities of water and fossil fuels (Pimentel et al. 2002), along with the utilization of pesticides and fertilizers for optimal production.

Another significant concern regarding soybean imports is the proliferation of transgenic varieties (Bonny 2008). Genetically modified crops pose threats to biodiversity conservation and sustainable agriculture, potentially resulting in the widespread contamination of natural flora with transgenic traits and the degradation and erosion of genetic resources. This concern is exacerbated in genetically modified plants engineered with pharmaceutical and industrial traits, as their release into the environment could have detrimental consequences (Garcia and Altieri 2005). Furthermore, the use of genetically modified organisms (GMOs) is prohibited in organic farming practices (Lombardo and Zelasco 2016), as organic regulations strictly prohibit such techniques (EU 848 2018).

In addition to these environmental and ethical concerns, there are apprehensions regarding the ethical implications of intense competition for arable land and non-renewable resources (such as water, fossil fuels, or minerals), which could lead to land grabbing by large-scale producers at the expense of small-scale farmers (Flachowsky et al. 2017).

Alternative legumes, including peas, chickpeas, lentils, and beans, are available as protein sources apart from soybeans. Although the cultivation of these legumes aligns with the principles of organic production by promoting short distribution circuits and local productions, their protein content falls significantly below that of soybean meals (Rudolph et al. 1983; Boye et al. 2010; EU 848 2018). Additionally, it's important to note that legumes contain some anti-nutritional factors that can inhibit the availability of essential nutrients. However, these factors can be mitigated or eliminated through various processing methods such as heating, enzyme application, soaking, germination, irradiation, fermentation, or mechanical methods like shelling and milling. Other techniques such as high-pressure processing, microwave heating, and extrusion can also help reduce these anti-nutritional factors (Nikmaram et al. 2017; Abbas and Ahmad 2018). However, implementing these processes adds to the labor and cost of utilizing them in feed production.

The aforementioned limitations underscore significant challenges in the use of protein sources for livestock feed within organic production systems. Therefore, there's a critical need to explore alternative solutions to overcome these obstacles.

5.4 Use of Insects as Alternative Protein Source in Animal Feed

Currently, insects are emerging as a new protein source for animal feed due to their high protein content, positioning themselves as a clear alternative to the dominant use of soybean meal (Biasato et al. 2016; Nieto et al. 2023). The class *Insecta* encompasses the largest variety of species globally, many of which offer abundant protein and amino acid content that can be effectively harnessed (Zou et al. 2024). Notably, black soldier flies, house flies, beetles, mealworms, silkworms, earthworms, crickets, and grasshoppers are gaining attention as they have been identified as significant future feed sources, meeting nutritional requirements, essential amino acid composition, nutrient digestibility, and feed acceptability (Belhadj Slimen et al. 2023).

These novel protein sources utilized in animal diets provide a compelling solution to mitigate the adverse social and ecological impacts stemming from extensive soybean imports, while also achieving comparable feed efficiency and product quality (Leiber et al. 2015). Insects address concerns associated with traditional protein sources by offering promising and sustainable prospects in their utilization (Sajid et al. 2023). This viable strategy, employing insects as a non-conventional protein source, contributes significantly to food security and environmental conservation (de Castro et al. 2018) (Fig. 5.2).

Among the advantages of insect meal production are its short life cycle, controllable cultivation, and lower requirements for natural resources such as water and land compared to conventional livestock farming (Sánchez-Muros et al. 2014; Perera and Bhujel 2021). Moreover, insects can be nourished with low-quality organic waste, alleviating pressure on crops intended for animal feed (Veldkamp and Bosch 2015). These attributes enable insects to contribute to reducing the

Fig. 5.2 Tenebrio molitor larvae meal

carbon footprint due to their efficient food conversion and decreased greenhouse gas emissions associated with their production (Vauterin et al. 2021). Therefore, their utilization not only promotes resource efficiency but also aids in mitigating the environmental impacts linked with intensive agriculture.

Insect meals possess acceptable nutritional properties, including an adequate amino acid profile (Sánchez-Muros et al., 2014). Their incorporation into feed formulations has been evaluated in various studies involving different species, yielding satisfactory results without adverse effects on growth and health status (Khan 2018; Perera and Bhujel 2021). Furthermore, insect meal usage has not compromised meat quality characteristics (Dabbou et al. 2020), and has even shown potential for enhancing results (Khan 2018; Perera and Bhujel 2021).

Currently, insect production is limited and costly. However, advancements in insect production technology and efficiency have the potential to scale up production while reducing costs, rendering insect meals more economically viable compared to conventional protein sources (Sajid et al. 2023).

The cultivation of insects for feed carries inherent biological risks and hazards, as bacteria, viruses, fungi, parasites, prions, and chemical contaminants may be present in unprocessed insects, depending on the substrate used for feeding and the rearing and harvesting conditions. However, the prevalence of some of these pathogens tends to be lower in insects compared to other sources of unprocessed animal proteins, as replication of certain pathogens within insects does not seem to occur (EFSA 2015).

Despite these risks, the use of insects in animal feed presents an encouraging option due to their nutritional qualities, which have shown positive effects on both yields and meat quality. However, the ultimate success hinges on consumer acceptance of meat obtained from animals fed with insect meal, as insects often evoke feelings of aversion and antipathy. To enhance consumer acceptance, it is crucial to raise awareness among consumers regarding the various benefits of incorporating insect meal into animal feed (Szendrő et al. 2020).

5.5 Regulation on the Inclusion of Insects in Organic Animal Feed

As previously mentioned, numerous studies have explored the effects of incorporating insects into the diets of various animal species, focusing on yields, quality, and environmental impact. However, a crucial aspect for the widespread adoption of insect-based feed lies in legislation, regulations, and food standards governing their use, which need to be developed further to facilitate the growth of the insect industry (Zou et al. 2024).

Organic livestock farming prioritizes extensive production systems where animals have access to open-air environments, a practice closely associated with animal welfare. Poultry, pigs, and fish naturally forage for insects when allowed to range freely, making insect meal a potentially acceptable substitute for soybean and fish meal as a protein source in their diets, without encountering consumer rejection (Szendrő et al. 2020).

In organic production, using insects as a form of biological control against pests is a common practice. This includes methods such as releasing sterile insects, using pheromones to monitor populations, and employing biopesticides (Baker et al. 2020). While the use of insects as feed for livestock is less prevalent due to modern regulations, Regulation (EU) 2018/848 of the European Parliament and the Council has expanded the definition of animal production to include insects (EU 848 2018). This development has paved the way for the implementation of insect-based feed in livestock production, offering a promising future for its use in the diets of animals raised under organic production regimes.

In its early stages, the EU began authorizing the use of farmed insects for animal feed as processed animal protein (PAP) in aquaculture production, compiling a list of approved insects (EU 893 2017). Insects have garnered increased commercial interest following the recent Commission Regulation (EU) 2021/1372 of 17 August 2021, which amended Annex IV to Regulation (EC) No. 999/2001 of the European Parliament and of the Council regarding the prohibition on feeding non-ruminant farmed animals, excluding fur animals, with proteins derived from animals. The use of processed animal protein is allowed in pigs, poultry, aquaculture animals, and fur animals, with the exception of ruminants. Animals from which such proteins are derived may feed any of the above-mentioned species except their own (EU 1372 2021).

These regulations establish the conditions for the production and use of insects for feed production for farmed animals other than fur animals. Insects are considered farm animals, and their breeding is classified under livestock farming. They can be utilized as raw material for feed, both dead and alive (with limitations). The insect species authorized for the production of hydrolyzed protein include the black soldier fly (*Hermetia illucens*), common house fly (*Musca domestica*), mealworm (*Tenebrio molitor*), bed beetle (*Alphitobius diaperinus*), domestic cricket (*Acheta domesticus*), striped cricket (*Gryllodes sigillatus*), and bicolor cricket (*Gryllus assimilis*). These insects may be imported from outside the Union if they correspond to the same species, the substrate for their feeding contains non-animal origin products or other listed products in the regulation, and if the feed has not come into contact with materials of non-permitted animal origin (EU 893 2017). Subsequently, the silkworm (*Bombyx mori*) has been added to this list of authorized species, and the term 'frass' has been included, which refers to the mixture of insect excrements with their dead parts or remains of the food substrate, with regulated conditions for its commercialization (EU 1925 2021). The nutritional characteristics of insects are influenced by various factors such as environmental conditions, feed composition, and species-specific traits (Bbosa et al. 2019). The main nutritional characteristics are listed in Table 5.1.

Some results obtained after its use in monogastrics and aquaculture are shown below:

Table 5.1 Main nutritional characteristics of the approved insect species

Insect	Crude protein (%)	Crude fat (%)
Hermetia illucens	42	29
Musca domestica	53.3	20.3
Tenebrio molitor	54.9	21.2
Alphitobius diaperinus	68	21
Acheta domesticus, Gryllus assimilis and *Gryllodes sigillatus*	59.5	22.6
Bombyx mori	55.6	32.2

(a) Black Soldier fly (*Hermetia illucens*)

The larvae of the black soldier fly are among the most commonly utilized insects in the diets of monogastric and aquaculture species, alongside those of *Tenebrio molitor*. Possess the capability to convert waste into feed, can be cultivated in simple facilities, and do not accumulate pesticides or mycotoxins (Wang and Shelomi 2017).

Black soldier fly meal has frequently been employed in broilers as a partial or full substitute for soybean meal, yielding no adverse effects on animal growth and health. Utilizing fat derived from these larvae has ensured satisfactory productive performances, carcass characteristics, and overall poultry meat quality (Schiavone et al. 2017; Bellezza Oddon et al. 2021), while also presenting lower risks for cardiovascular health (Mlaga et al. 2022). Its incorporation into organic chicken diets has likewise not compromised yields or meat quality (Leiber et al. 2015). In some instances, a partial substitution is deemed more appropriate than using high amounts of the insect, both in terms of developmental outcomes and product quality (Murawska et al. 2021).

The administration of larvae to fattening pigs has yielded beneficial impacts on growth performance and meat quality (Yu et al. 2019), and has also enhanced piglet development and behavior by reducing post-weaning stress when supplied live (Ipema et al. 2021). Furthermore, its inclusion in aquaculture species diets has been shown to enhance performance by bolstering immunity and disease resistance (Mohan et al. 2022).

(b) Common house fly (*Musca domestica*)

The nutritional analysis of house fly larvae reveals a high protein content and a favorable amino acid profile, particularly rich in essential amino acids such as lysine and methionine. Studies have demonstrated successful outcomes when incorporating these larvae into broiler diets (Hall et al. 2018). Their inclusion in feed formulations has led to increased body weight gain without adverse effects on carcass characteristics (Reda et al. 2023). They exhibit significant potential as a protein supplement for commercial turkey diets, particularly in the context of poultry farming (Zuidhof et al. 2003).

While their utilization in swine diets is less common, house fly larvae have been more frequently employed for manure treatment (Roffeis et al. 2015). Nevertheless, they are also regarded as an alternative protein source in swine nutrition (Hong and Kim 2022).

In aquaculture, house fly larvae have been utilized effectively, without inducing physiological stress in fish, and have often led to improved yields. This positions worm meals as a feed with high nutritional quality, owing to their rich content in proteins, fats, minerals, B-complex vitamins, trace elements, and phosphorus. Their high acceptance by fish and crustaceans underscores their value in aquaculture settings (De Leon Ramirez et al. 2018; Saleh 2020).

(c) Yellow mealworm (*Tenebrio molitor*)

The utilization of mealworm larvae, in its coleopteran larval form, ranks among the most common raw materials alongside larvae of the black soldier fly. Its high nutritional value is comparable to that of soybean and fish meal, with calcium being the most limiting nutrient, easily supplemented in the diet (Selaledi et al. 2020; Nieto et al. 2022). The poultry sector has been one of the most extensively evaluated areas, with the inclusion of mealworm larvae meals in broiler diets resulting in benefits such as improved body weight gain and feed conversion rates (Sedgh-Gooya et al. 2021).

Various studies have examined different percentages of substitution of soybean meals with mealworm larvae meals, with findings suggesting that high inclusion percentages are not always recommended. Lower inclusions have been found to favor meat quality in broilers (Shaviklo et al. 2021), patridges (Secci et al. 2018) and quail (Secci et al. 2022).

In free-range chickens, as observed in organic production systems, the use of mealworm larvae meals has not negatively impacted welfare, productive performance, or carcass characteristics, nor has it affected the physical and chemical properties of meat (Biasato et al. 2016; Dabbou et al. 2020). Even with full substitutions, satisfactory outcomes have been reported (Bovera et al. 2016). The substitution of soybean meal with mealworm larvae meal, beginning from the first day of life until slaughter in free-range chickens, suggests partial substitutions during the first month and complete substitution as the animals mature, yielding satisfactory results (Nieto et al. 2022).

While its use in weaned pigs hasn't shown adverse effects on performance, complete protein substitutions are not recommended (Ko et al. 2020; Hong and Kim 2022). In aquaculture species, mealworm larvae meals have demonstrated no detrimental effects on nutrient digestibility and growth performance, while ensuring nutritional value (Basto et al. 2021). These findings position mealworm larvae as promising candidates for new protein sources in aquafeeds without compromising yields (Chemello et al. 2020).

(d) Bed beetle (*Alphitobius diaperinus*)

The larvae of this coleopteran have a wide variety of amino acids (Despins and Axtell 1995). Interestingly, the pupae of this insect are characterized by lower chitin content and may contain more protein and less fat compared to larvae, rendering their utilization in pupal form a promising option for animal feed (Kurečka et al. 2021).

While the use of this coleopteran in animal feed has been less widespread compared to other insects, notable findings have emerged. For instance, employing litter beetle meal as a substitute for soybean meal in piglets did not adversely affect growth performance, carcass composition, or meat quality, although it did result in a higher content of polyunsaturated fatty acids (PUFA) in adipose tissue (Müller Richli et al. 2023).

In aquaculture, when used as a substitute for fish meal, it positively influenced the growth, health status, and sensory attributes of European perch (Tran et al. 2024). However, its use in broiler chicks did not yield similar positive outcomes (Despins and Axtell 1995).

(e) Domestic cricket (*Acheta domesticus*), bicolor cricket (*Gryllus assimilis*) and striped cricket (*Gryllodes sigillatus*)

The three insect species belonging to the family *Gryllidae* do not undergo a larval stage as they exhibit incomplete metamorphosis. Their crude protein and fat content make them a viable alternative for inclusion in animal diets (DeFoliart et al. 1982; Wang et al. 2008; Nieto et al. 2023). Domestic crickets, known for their digestibility of dry matter, organic matter, and crude protein, hold significant potential for utilization in feed formulations (Kovitvadhi et al. 2019).

Studies have shown that the incorporation of adult crickets into broiler diets does not hinder animal growth or alter meat flavor (Finke et al. 1985; Nakagaki et al. 1987). In fact, certain investigations have demonstrated improved growth compared to conventional diets (DeFoliart et al. 1982). Partial incorporations of cricket meal in broiler diets during the early stages of life have yielded better outcomes than complete substitution of conventional feeds (Wang et al. 2005). According to Nieto et al. (2023), complete substitution of soybean meal with *Acheta domesticus* meal in free-range slow-growing broilers yielded favorable results throughout their growth cycle, except during the first weeks of life, where partial substitution is recommended.

In aquaculture, partial substitution of fish meal with cricket meal has demonstrated superior results compared to total substitution when feeding Nile Tilapia fry (Perera and Bhujel 2021). Additionally, in weaned pigs, field cricket as a partial substitute for soybean and fish meal exhibited beneficial effects on growth, gut morphology, antioxidant capacity, and gut microbiota (Boontiam et al. 2022).

(f) Silkworm (*Bombyx mori*)

The silkworm was the most recent addition to the list of insects authorized for the production of processed animal proteins for the manufacture of feed for farm animals (EU 1925 2021). Silkworm pupae have a high protein and lipid content, as well as high levels of essential amino acids (Tomotake et al. 2010).

These pupae present an intriguing resource for animal feed, including aquaculture. Studies have shown that their incorporation into extruded fish feed up to 15%

did not compromise its physical quality (Alcaraz et al. 2021). In grower diets for broilers, soybean meal can be effectively replaced with silkworms without compromising production and slaughter performance (Ullah et al. 2017; Pietras et al. 2021), with some studies even indicating improvements in quality indices (Zhao et al. 2023).

Furthermore, in fattening pigs, the inclusion of silkworm pupa meal in small percentages has been found to enhance growth, nutrient utilization, and feed conversion efficiency, while also reducing feed costs (Barman et al. 2021).

5.6 Conclusions

In conclusion, insect meals emerge as a compelling alternative to traditional protein sources in organic livestock farming, offering a sustainable solution without compromising production outcomes, animal welfare, or meat quality. Partial substitutions in poultry, swine, and aquaculture have demonstrated favorable results, with total substitutions being more suitable as animals mature, preferably avoiding the early stages of life. Therefore, integrating insects into organic production, in carefully calibrated quantities tailored to the animal's age, holds the potential to enhance both production efficiency and meat quality. This underscores the notion that organic farming practices can yield products of superior quality compared to conventional methods.

References

Abbas Y, Ahmad A (2018) Impact of processing on nutritional and antinutritional factors of legumes: a review. Ann Food Sci Technol 19:1–17

Abro Z, Kassie M, Tanga C, Beesigamukama D, Diiro G (2020) Socio-economic and environmental implications of replacing conventional poultry feed with insect-based feed in Kenya. J Clean Prod 265:121871. https://linkinghub.elsevier.com/retrieve/pii/S0959652620319181

Alcaraz R, Mita GI, Hernández-Contreras A, Hernández MD (2021) Physical properties of extruded fish feed containing silkworm (Bombyx mori) pupae meal. J Insects Food Feed 7:449–456

Allegretti G, Talamini E, Schmidt V, Bogorni PC, Ortega E (2018) Insect as feed: An emergy assessment of insect meal as a sustainable protein source for the Brazilian poultry industry. J Clean Prod 171:403–412. https://linkinghub.elsevier.com/retrieve/pii/S0959652617322564

Baker BP, Green TA, Loker AJ (2020) Biological control and integrated pest management in organic and conventional systems. Biol Control 140:104095

Barman K, Banik S, Thomas R, Kumar S, Das AK, Dutta K, Rajkhowa S (2021) Effect of partial replacement of protein supplement with silkworm (Bombyx mori L.) pupa meal on production performances in crossbred (HS × GH) grower pigs. Indian J Anim Sci 90:1519–1523

Basto A, Calduch-Giner J, Oliveira B, Petit L, Sá T, Maia MRG, Fonseca SC, Matos E, Pérez-Sánchez J, Valente LMP (2021) The use of defatted Tenebrio molitor larvae meal as a Main protein source is supported in European Sea bass (Dicentrarchus labrax) by data on growth performance, lipid metabolism, and flesh quality. Front Physiol 12. https://www.frontiersin.org/articles/10.3389/fphys.2021.659567/full

Bbosa T, Tamale Ndagire C, Muzira Mukisa I, Fiaboe KKM, Nakimbugwe D (2019) Nutritional characteristics of selected insects in Uganda for use as alternative protein sources in food and feed (P Weintraub, Ed.). J Insect Sci 19. https://academic.oup.com/jinsectscience/article/doi/10.1093/jisesa/iez124/5680619

Belhadj Slimen I, Yerou H, Ben Larbi M, M'Hamdi N, Najar T (2023) Insects as an alternative protein source for poultry nutrition: a review. Front Vet Sci 10

Bellezza Oddon S, Biasato I, Imarisio A, Pipan M, Dekleva D, Colombino E, Capucchio MT, Meneguz M, Bergagna S, Barbero R, Gariglio M, Dabbou S, Fiorilla E, Gasco L, Schiavone A (2021) Black soldier fly and yellow mealworm live larvae for broiler chickens: effects on bird performance and health status. J Anim Physiol Anim Nutr (Berl) 105:10–18. https://onlinelibrary.wiley.com/doi/full/10.1111/jpn.13567 (verified 20 January 2022)

Biasato I, De Marco M, Rotolo L, Renna M, Lussiana C, Dabbou S, Capucchio MT, Biasibetti E, Costa P, Gai F, Pozzo L, Dezzutto D, Bergagna S, Martínez S, Tarantola M, Gasco L, Schiavone A (2016) Effects of dietary Tenebrio molitor meal inclusion in free-range chickens. J Anim Physiol Anim Nutr (Berl) 100:1104–1112. http://doi.wiley.com/10.1111/jpn.12487

Bonny S (2008) Genetically modified glyphosate-tolerant soybean in the USA: adoption factors, impacts and prospects. A review. Agron Sustain Dev 28:21–32

Boontiam W, Hong J, Kitipongpysan S, Wattanachai S (2022) Full-fat field cricket (Gryllus bimaculatus) as a substitute for fish meal and soybean meal for weaning piglets: effects on growth performance, intestinal health, and redox status. J Anim Sci 100

Bovera F, Loponte R, Marono S, Piccolo G, Parisi G, Iaconisi V, Gasco L, Nizza A (2016) Use of Tenebrio molitor larvae meal as protein source in broiler diet: effect on growth performance, nutrient digestibility, and carcass and meat traits. J Anim Sci 94:639–647. https://academic.oup.com/jas/article/94/2/639/4701579

Boye J, Zare F, Pletch A (2010) Pulse proteins: processing, characterization, functional properties and applications in food and feed. Food Res Int 43:414–431. https://linkinghub.elsevier.com/retrieve/pii/S0963996909002683

Chemello G, Renna M, Caimi C, Guerreiro I, Oliva-Teles A, Enes P, Biasato I, Schiavone A, Gai F, Gasco L (2020) Partially defatted Tenebrio molitor Larva meal in diets for grow-out rainbow trout, Oncorhynchus mykiss (Walbaum): effects on growth performance, diet digestibility and metabolic responses. Animals 10:229

Dabbou S, Gasco L, Lussiana C, Brugiapaglia A, Biasato I, Renna M, Cavallarin L, Gai F, Schiavone A (2020) Yellow mealworm (Tenebrio molitor L.) larvae inclusion in diets for free-range chickens: effects on meat quality and fatty acid profile. Renew Agric Food Syst 35:571–578. https://www.cambridge.org/core/product/identifier/S1742170519000206/type/journal_article

de Castro RJS, Ohara A, dos S. Aguilar JG, Domingues MAF (2018) Nutritional, functional and biological properties of insect proteins: processes for obtaining, consumption and future challenges. Trends Food Sci Technol 76:82–89. https://linkinghub.elsevier.com/retrieve/pii/S0924224417306702 (verified 23 June 2022)

De Leon Ramirez J, Garcia Trejo JF, Sosa Ferreyra C, Martinez Ramos S, Bottini Cedeno B (2018) Fly larvae (Musca domestica) as a protein alternative in the feeding of Macrobrachium tenellum. Lat Am J Aquat Res 46:599–603

de Souza J, Andronicos NM, Kolakshyapati M, Hilliar M, Sibanda TZ, Andrew NR, Swick RA, Wilkinson S, Ruhnke I (2021) Black soldier fly larvae in broiler diets improve broiler performance and modulate the immune system. Animal Nutr 7:695–706. https://linkinghub.elsevier.com/retrieve/pii/S2405654521000871

DeFoliart GR, Finke MD, Sunde ML (1982) Potential value of the Mormon cricket (Orthoptera: Tettigoniidae) harvested as a high-protein feed for poultry. J Econ Entomol 75:848–852. https://academic.oup.com/jee/article/75/5/848/2213654 (verified 26 January 2022)

Despins JL, Axtell RC (1995) Feeding behavior and growth of broiler chicks fed larvae of the darkling beetle, Alphitobius diaperinus. Poult Sci 74:331–336. https://linkinghub.elsevier.com/retrieve/pii/S0032579119453637

EFSA (2015) Risk profile related to production and consumption of insects as food and feed. EFSA J 13:4257
Escribano A (2018) Organic feed: A bottleneck for the development of the livestock sector and its transition to sustainability? Sustain For 10:2393
EU 1372 (2021) Commission Regulation (EU) 2021/1372 of 17 August 2021 amending Annex IV to Regulation (EC) No 999/2001 of the European Parliament and of the Council as regards the prohibition to feed non-ruminant farmed animals, other than fur animals, with protein derived from animals
EU 1925 (2021) Commission Regulation (EU) 2021/1925 of 5 November 2021 amending certain Annexes to Regulation (EU) No 142/2011 as regards the requirements for placing on the market of certain insect products and the adaptation of a containment method
EU 848 (2018) Regulation (EU) 2018/848 of the European Parliament and of the Council of 30 May 2018 on organic production and labelling of organic products and repealing Council Regulation (EC) No 834/2007. https://eur-lex.europa.eu/legal-content/ES/TXT/PDF/?uri=CELEX:32018R0848&from=EN
EU 893 (2017) Commission Regulation (EU) 2017/893 of 24 May 2017 amending Annexes I and IV to Regulation (EC) No 999/2001 of the European Parliament and of the Council and Annexes X, XIV and XV to Commission Regulation (EU) No 142/2011 as regards the provisions on processed animal protein (Text with EEA relevance.)
Fehlenberg V, Baumann M, Gasparri NI, Piquer-Rodriguez M, Gavier-Pizarro G, Kuemmerle T (2017) The role of soybean production as an underlying driver of deforestation in the South American Chaco. Glob Environ Chang 45:24–34
Finke MD, Sunde ML, Defoliart GR (1985) An evaluation of the protein quality of Mormon crickets (Anabrus simplex Haldeman) when used as a high protein feedstuff for poultry. Poult Sci 64:708–712. https://linkinghub.elsevier.com/retrieve/pii/S0032579119349934 (verified 26 January 2022)
Flachowsky G, Meyer U, Südekum K-H (2017) Land use for edible protein of animal origin—a review. Animals 7:25
Garcia MA, Altieri MA (2005) Transgenic crops: implications for biodiversity and sustainable agriculture. Bull Sci Technol Soc 25:335–353
Hall HN, Masey O'Neill HV, Scholey D, Burton E, Dickinson M, Fitches EC (2018) Amino acid digestibility of larval meal (Musca domestica) for broiler chickens. Poult Sci 97:1290–1297. https://linkinghub.elsevier.com/retrieve/pii/S0032579119309952
Hong J, Kim YY (2022) Insect as feed ingredients for pigs. Anim Biosci 35:347–355. http://animbiosci.org/journal/view.php?doi=10.5713/ab.21.0475
Hovhannisyan SV, Grigoryan KA (2016) The main problems and features of the global and local meat production. Ann Agrar Sci 14:315–318
Ipema AF, Bokkers EAM, Gerrits WJJ, Kemp B, Bolhuis JE (2021) Providing live black soldier fly larvae (Hermetia illucens) improves welfare while maintaining performance of piglets post-weaning. Sci Rep 11:7371. https://www.nature.com/articles/s41598-021-86765-3
Karlsson JO, Parodi A, van Zanten HHE, Hansson P-A, Röös E (2020) Halting European Union soybean feed imports favours ruminants over pigs and poultry. Nat Food 2:38–46
Khan SH (2018) Recent advances in role of insects as alternative protein source in poultry nutrition. J Appl Anim Res 46:1144–1157. https://www.tandfonline.com/doi/abs/10.1080/09712119.2018.1474743 (verified 13 January 2022)
Khan M, Chand N, Khan S, Naz S, Alrefaei AF, Almutairi MH, Sifa D, Khan RU (2023) Partial and total replacement of soybean meal with maggot larvae meal in broiler chicken diets: impact on growth performance, nutrients digestibility, haemotological parameters and intestinal histomorphology. Livest Sci 277:105355. https://linkinghub.elsevier.com/retrieve/pii/S1871141323002019
Kim SW, Less JF, Wang L, Yan T, Kiron V, Kaushik SJ, Lei XG (2019) Meeting global feed protein demand: challenge, opportunity, and strategy. Annu Rev Anim Biosci 7:221–243. https://www.annualreviews.org/doi/10.1146/annurev-animal-030117-014838

Ko H, Kim Y, Kim J (2020) The produced mealworm meal through organic wastes as a sustainable protein source for weanling pigs. J Anim Sci Technol 62:365–373

Kovitvadhi A, Chundang P, Luapan J, Amarapitak P, Sriyaphai P, Buahom R, Cham-iam T, Leelehapongsathon K, Tirawattanawanich C, Thongprajukaew K (2019) Cricket powder as an alternative protein source for broilers based on in vitro digestibility. J Insects Food Feed 5:185–191. https://www.wageningenacademic.com/doi/10.3920/JIFF2018.0043 (verified 26 January 2022)

Kurečka M, Kulma M, Petříčková D, Plachý V, Kouřimská L (2021) Larvae and pupae of Alphitobius diaperinus as promising protein alternatives. Eur Food Res Technol 247:2527–2532

Kushwah, S., Dhir, A., Sagar, M., & Gupta, B. (2019). Determinants of organic food consumption. A systematic literature review on motives and barriers. Appetite, 143, 104402. https://doi.org/10.1016/j.appet.2019.104402

Leiber F, Gelencsér T, Stamer A, Amsler Z, Wohlfahrt J, Früh B, Maurer V (2015) Insect and legume-based protein sources to replace soybean cake in an organic broiler diet: effects on growth performance and physical meat quality. Renew Agric Food Syst 32:21–27. https://www.cambridge.org/core/product/identifier/S1742170515000496/type/journal_article

Lombardo L, Zelasco S (2016) Biotech approaches to overcome the limitations of using transgenic plants in organic farming. Sustain For 8:497

Mlaga KG, Agboka K, Attivi K, Tona K, Osseyi E (2022) Assessment of the chemical characteristics and nutritional quality of meat from broiler chicken fed black soldier fly (Hermetia illucens) larvae meal. Heliyon:e11718. https://linkinghub.elsevier.com/retrieve/pii/S2405844022030067

Mohan K, Rajan DK, Muralisankar T, Ganesan AR, Sathishkumar P, Revathi N (2022) Use of black soldier fly (Hermetia illucens L.) larvae meal in aquafeeds for a sustainable aquaculture industry: A review of past and future needs. Aquaculture 553:738095

Müller Richli M, Weinlaender F, Wallner M, Pöllinger-Zierler B, Kern J, Scheeder MRL (2023) Effect of feeding Alphitobius diaperinus meal on fattening performance and meat quality of growing-finishing pigs. J Appl Anim Res 51:204–211

Murawska D, Daszkiewicz T, Sobotka W, Gesek M, Witkowska D, Matusevičius P, Bakuła T (2021) Partial and Total replacement of soybean meal with full-fat black soldier Fly (Hermetia illucens L.) larvae meal in broiler chicken diets: impact on growth performance, carcass quality and meat quality. Animals 11:2715. https://www.mdpi.com/2076-2615/11/9/2715

Nakagaki BJ, Sunde ML, Defoliart GR (1987) Protein quality of the house cricket, Acheta domesticus, when fed to broiler chicks. Poult Sci 66:1367–1371. https://linkinghub.elsevier.com/retrieve/pii/S0032579119563155 (verified 26 January 2022)

Nieto J, Plaza J, Lara J, Abecia J-A, Revilla I, Palacios C (2022) Performance of slow-growing chickens fed with Tenebrio molitor larval meal as a full replacement for soybean meal. Vet Sci 9:131. https://www.mdpi.com/2306-7381/9/3/131

Nieto J, Plaza J, Lara J, Abecia J-A, Revilla I, Palacios C (2023) Use of Acheta domesticus meal as a full soybean substitute in the feeding of slow-growing chicks. Poult Sci 102:102503. https://linkinghub.elsevier.com/retrieve/pii/S0032579123000299

Nikmaram N, Leong SY, Koubaa M, Zhu Z, Barba FJ, Greiner R, Oey I, Roohinejad S (2017) Effect of extrusion on the anti-nutritional factors of food products: an overview. Food Control 79:62–73

Perera AD, Bhujel RC (2021) Field cricket (Gryllus bimaculatus) meal (FCM) to replace fishmeal in the diets for sex reversal and nursing of Nile tilapia (Oreochromis niloticus) fry. Aquac Res 52:4946–4958. https://onlinelibrary.wiley.com/doi/10.1111/are.15328

Pietras M, Orczewska-Dudek S, Szczurek W, Pieszka M (2021) Effect of dietary lupine seeds (Lupinus luteus L.) and different insect larvae meals as protein sources in broiler chicken diet on growth performance, carcass, and meat quality. Livest Sci 250:104537. https://linkinghub.elsevier.com/retrieve/pii/S1871141321001451

Pimentel D, Doughty R, Carothers C, Lamberson S, Bora N, Lee K (2002) Energy inputs in crop production in developing and developed countries. Food security and environmental quality in the developing world, vol 2. CRC Press, pp 129–151

Reda AA, Hussien FZ, Gizaw TT, Beza GM (2023) Research note: effects of partial replacement of noug seed (Guizotia abyssinica) cake with maggot meal in dual-purpose chicken. Poult Sci 102:102760. https://linkinghub.elsevier.com/retrieve/pii/S0032579123002791

Richards PD, Myers RJ, Swinton SM, Walker RT (2012) Exchange rates, soybean supply response, and deforestation in South America. Glob Environ Chang 22:454–462

Roffeis M, Muys B, Almeida J, Mathijs E, Achten WMJ, Pastor B, Velásquez Y, Martinez-Sanchez AI, Rojo S (2015) Pig manure treatment with housefly (Musca domestica) rearing—an environmental life cycle assessment. J Insects Food Feed 1:195–214

Rudolph BC, Boggs LS, Knabe DA, Tanksley TD, Anderson SA (1983) Digestibility of nitrogen and amino acids in soybean products for pigs. J Anim Sci 57:373–386. https://academic.oup.com/jas/article/57/2/373-386/4744346

Ruiz-Mirazo J (2011) Environmental benefits of extensive livestock farming: wildfire prevention and beyond

Sajid QUA, Asghar MU, Tariq H, Wilk M, Płatek A (2023) Insect meal as an alternative to protein concentrates in poultry nutrition with future perspectives (An updated review). Agriculture 13:1239

Saleh HHE (2020) Review on using of housefly maggots (Musca domestica) in fish diets. J Zool Res 2

Sánchez-Muros M-J, Barroso FG, Manzano-Agugliaro F (2014) Insect meal as renewable source of food for animal feeding: a review. J Clean Prod 65:16–27. https://linkinghub.elsevier.com/retrieve/pii/S095965261300841X (verified 23 June 2022)

Schiavone A, Cullere M, De Marco M, Meneguz M, Biasato I, Bergagna S, Dezzutto D, Gai F, Dabbou S, Gasco L, Dalle Zotte A (2017) Partial or total replacement of soybean oil by black soldier fly larvae (Hermetia illucens L.) fat in broiler diets: effect on growth performances, feed-choice, blood traits, carcass characteristics and meat quality. Ital J Anim Sci 16:93–100. https://www.tandfonline.com/doi/full/10.1080/1828051X.2016.1249968

Secci G, Moniello G, Gasco L, Bovera F, Parisi G (2018) Barbary partridge meat quality as affected by Hermetia illucens and Tenebrio molitor larva meals in feeds. Food Res Int 112:291–298. https://linkinghub.elsevier.com/retrieve/pii/S0963996918304927

Secci G, Dabbou S, Lira de Medeiros AC, Addeo NF, Atallah E, Parisi G, Moniello G, Bovera F (2022) Low dietary inclusion levels of Tenebrio molitor larva meal slightly modify growth performance, carcass and meat traits of Japanese quail (Coturnix japonica). J Sci Food Agric 102:6578–6585. https://onlinelibrary.wiley.com/doi/10.1002/jsfa.12023

Sedgh-Gooya S, Torki M, Darbemamieh M, Khamisabadi H, Karimi Torshizi MA, Abdolmohamadi A (2021) Yellow mealworm, Tenebrio molitor (Col: Tenebrionidae), larvae powder as dietary protein sources for broiler chickens: effects on growth performance, carcass traits, selected intestinal microbiota and blood parameters. J Anim Physiol Anim Nutr (Berl) 105:119–128. https://onlinelibrary.wiley.com/doi/10.1111/jpn.13434

Selaledi L, Mbajiorgu CA, Mabelebele M (2020) The use of yellow mealworm (T. molitor) as alternative source of protein in poultry diets: a review. Trop Anim Health Prod 52:7–16. https://link.springer.com/article/10.1007/s11250-019-02033-7 (verified 20 January 2022)

Shaviklo AR, Alizadeh-Ghamsari AH, Hosseini SA (2021) Sensory attributes and meat quality of broiler chickens fed with mealworm (Tenebrio molitor). J Food Sci Technol 58:4587–4597. https://link.springer.com/10.1007/s13197-020-04946-w

Soisontes S, Freund F, Laquai V (2023) The future of protein feed: a case study of sustainable substitutes in the German livestock industry, vol 26. Environ Dev Sustain, pp 25199–25226

Sundrum A (2001) Organic livestock farming. Livest Prod Sci 67:207–215. https://linkinghub.elsevier.com/retrieve/pii/S0301622600001883

Szendrő K, Nagy MZ, Tóth K (2020) Consumer acceptance of meat from animals reared on insect meal as feed. Animals 10:1312

te Pas MFW, Veldkamp T, de Haas Y, Bannink A, Ellen ED (2021) Adaptation of livestock to new diets using feed components without competition with human edible protein sources—

a review of the possibilities and recommendations. Animals 11:2293. https://www.mdpi.com/2076-2615/11/8/2293
Thornton PK (2010) Livestock production: recent trends, future prospects. Philos Trans R Soc B Biol Sci 365:2853–2867
Tomotake H, Katagiri M, Yamato M (2010) Silkworm pupae (Bombyx mori) are new sources of high quality protein and lipid. J Nutr Sci Vitaminol (Tokyo) 56:446–448. http://www.jstage.jst.go.jp/article/jnsv/56/6/56_6_446/_article
Tran HQ, von Siebenthal EW, Luce J-B, Nguyen TT, Stejskal V, Weinlaender F, Janssens T (2024) A novel protein source from lesser mealworm (Alphitobius diaperinus) larvae meal for European perch (Perca fluviatilis): investigation on pellet characteristics, production performance, serum biochemistry, digestibility, histology, sensory and trait of fillet, and environmental indices. Aquaculture 581:740460
Ullah R, Khan S, Hafeez A, Sultan A, Khan NA, Chand N, Ahmad N (2017) Silkworm (Bombyx mori) meal as alternate protein ingredient in broiler finisher ration. Pak J Zool 49:1463–1470
Vauterin A, Steiner B, Sillman J, Kahiluoto H (2021) The potential of insect protein to reduce food-based carbon footprints in Europe: the case of broiler meat production. J Clean Prod 320:128799. https://linkinghub.elsevier.com/retrieve/pii/S0959652621029978
Veldkamp T, Bosch G (2015) Insects: a protein-rich feed ingredient in pig and poultry diets. Animal. Frontiers 5:45–50. https://www.researchgate.net/publication/274890036_Insects_a_protein-rich_feed_ingredient_in_pig_and_poultry_diets (verified 23 June 2022)
Wang Y-S, Shelomi M (2017) Review of black soldier Fly (Hermetia illucens) as animal feed and human food. Food Secur 6:91. http://www.mdpi.com/2304-8158/6/10/91
Wang D, Zhai SW, Zhang CX, Bai YY, An SH, Xu YN (2005) Evaluation on nutritional value of field crickets as a poultry feedstuff. Asian-Australas J Anim Sci 18:667–670. http://ajas.info/journal/view.php?doi=10.5713/ajas.2005.667 (verified 26 January 2022)
Wang D, Bai YY, Li JH, Zhang CX (2008) Nutritional value of the field cricket (Gryllus testaceus walker). Insect Sci 11:275–283. https://onlinelibrary.wiley.com/doi/full/10.1111/j.1744-7917.2004.tb00424.x (verified 26 January 2022)
Yu M, Li Z, Chen W, Rong T, Wang G, Li J, Ma X (2019) Use of Hermetia illucens larvae as a dietary protein source: effects on growth performance, carcass traits, and meat quality in finishing pigs. Meat Sci 158:107837
Zhao X, Kuang Z, Yang J, Bilal M, Yang R (2023) Effect of fermented silkworm (Bombyx mori) pupae on growth performance, immune index and meat quality in Lingnan yellow broiler chickens. Pak J Zool 55. http://researcherslinks.com/current-issues/Effect-of-Fermented-Silkworm-Bombyx-mori-Pupae-on-Growth/20/1/6410/html
Zou X, Liu M, Li X, Pan F, Wu X, Fang X, Zhou F, Peng W, Tian W (2024) Applications of insect nutrition resources in animal production. J Agric Food Res 15:100966. https://linkinghub.elsevier.com/retrieve/pii/S2666154324000036
Zuidhof MJ, Molnar CL, Morley FM, Wray TL, Robinson FE, Khan BA, Al-Ani L, Goonewardene LA (2003) Nutritive value of house fly (Musca domestica) larvae as a feed supplement for Turkey poults. Anim Feed Sci Technol 105:225–230. https://linkinghub.elsevier.com/retrieve/pii/S037784010300004X

Chapter 6
Comparison of Milk Yield and Milk Composition Between Organic and Conventional Systems: A Meta-analysis and Meta-regression

Juan Carlos Angeles-Hernandez, Rodolfo Vieyra Alberto, Ana Karen Zaldivar-Ortega, Navid Ghavipanje, Alfredo J. Escribano, and Manuel Gonzalez Ronquillo

Abstract There is growing consumers concern regarding consumption of food products that not only provide the main nutrients, but also provide of nutraceutical proprieties. In this sense, the organic products are increasingly favored as an option to get a healthy food consumption. The aim of this study was to compare conventional system (CS) and organic system (OS) dairy production using a systematic review and meta-analysis approach. An analytical dataset was constructed from 214 different studies. Results showed that animals managed under OS produce less milk (SMD −0.65; 95% CI; $p = 0.0007$) compared to those under CS. There were no significant differences in fat, protein, lactose, and monounsaturated fatty acids (MUFA) concentrations between OS and CS milk. However, OS milk had higher polyunsaturated fatty acids (PUFA; SMD +1.13; 95% CI; $p = 0.005$) and

J. C. Angeles-Hernandez (✉)
Departamento de Medicina y Zootecnia de Rumiantes, Facultad de Medicina Veterinaria y Zootecnia, Universidad Nacional Autónoma de México, Mexico City, CP, Mexico
e-mail: juanangeles@fmvz.unam.mx

R. Vieyra Alberto · A. K. Zaldivar-Ortega
Instituto de Ciencias Agropecuarias, Universidad Autónoma del Estado de Hidalgo, Tulancingo de Bravo, Hidalgo, Mexico

N. Ghavipanje
Department of Animal Science, Faculty of Agriculture, University of Birjand, Birjand, Iran

South Khorasan Agricultural and Natural Resources Research and Education Center, Agricultural Research, Education and Extension Organization (AREEO), Birjand, Iran

A. J. Escribano
Independent Researcher & Consultant, Cáceres, Spain

M. Gonzalez Ronquillo
Facultad de Medicina Veterinaria y Zootecnia, Universidad Autonoma del Estado de Mexico, Toluca, Mexico
e-mail: mrg@uaemex.mx

C. Palacios Riocerezo, M. Gonzalez Ronquillo (eds.), *Traditional Livestock Production*, https://doi.org/10.1007/978-3-031-88275-3_6

alpha-linolenic acid (ALA; SMD +1.54; 95% CI; $p = 0\ .0004$) both accompanied by high heterogeneity ($I^2 > 93.0\%$). Also, we found that the content of conjugated linoleic acid (CLA) was higher in OS (SMD +3.06; 95% CI; $p = 0.04$) with a large heterogeneity ($I^2 > 94.0\%$). We conclude that based on the analyzed parameters, mainly the FA profile, the milk from OS can be a healthy option for human consumption. However, other milk characteristics such as residues (e.g. antibiotics) need to be considered in future studies to compare CS and OS systems. Finally, the identified differences between CS and OS production systems can mainly be explained by the animal species and the feeding management.

Keywords Dairy production system · Fatty acids · Meta-analysis · Milk composition · Organic products

6.1 Introduction

Organic food has been recognized by consumers as a trend for 'healthier' products besides promoting animal- and eco-friendly production practices (Vargas-Bello-Pérez et al. 2020; Linehan et al. 2024). Global sales of organic products reached 135 billion euros in 2022, and it has proven to be highly relevant in different countries (FAO 2024). Organic dairy products account for almost 30% of the total milk market (Bórawski et al. 2021). Global milk production reached 893 billion liters in 2019 and is projected to grow by 22% by 2027 (Bórawski et al. 2021; Linehan et al. 2024). According to the FAO, in 2019 the European Union had a production of 156.2 billion, followed by the United States with a production of 99 billion milk (Gutierrez et al. 2022). Of this milk production, 87.7% is cow's milk and 1.8% is produced in organic production systems (Bórawski et al. 2021).

In the food industry, milk and dairy products are important sources of macro- and micro-nutrients, including high-quality proteins, fats, calcium, potassium, phosphorus, vitamin D, riboflavin, and vitamin B12 for human (Linehan et al. 2024). In addition, the milk has been considered as a food-functional due to the presence of probiotic compound and contains viable bacteria which are beneficial to human body (Akram et al. 2020). The majority of global milk production carried out within conventional systems (CS, i.e. traditional and intensive systems) that typically prioritize high productivity; however, organic systems (OS) production is in a period of rapid development (Brodziak et al. 2021). Milk production in CS heavily relies on mechanization, synthetics, and antibiotics, leading to environmental damage, animal welfare issues, and health risks (Linehan et al. 2024). To address this, eco- and animal-friendly practices, such as OS milk farming has been gained attention (Manuelian et al. 2020). According to the European Commission regulation (No. 889/2008 and No. 848/2018), OS cattle have higher access to organic grassland (untreated pesticides and synthetic fertilizers), more outdoor area than conventional and roughage should comprise the largest part of their diets. Consumers are willing to pay more for these dairy products (Padel et al. 2010).

It has been well established that different milk production managements (i.e. CS and OS) could affect not only quality of products but also environmental sustainability (Mie et al. 2017). In this sense, CS milk production has been found to be more efficient in terms of feed conversion, which leads to lower intensity of emissions (CO_2 equivalents per kg of product), having a contribution of 48–65% to the global warming through greenhouse gas (GHG) emissions (De Boer 2003). Under organic systems, carbon footprint is usually lower, especially when the limit of the study is the farm, instead of the approach of intensity of emissions. In this sense, Lambotte et al. (2023), analyzed 3074 French dairy farms and found that the carbon footprint of organic milk is between 19% and 11% lower than its conventional counterpart. Differences are greatly influenced by the land use change, the production system design, and therefore, when carbon sequestration is considered. This is especially noticeable in complex agro-ecosystems, such as agroforestry, where carbon sequestration is higher, therefore carbon balance shows better environmental performance in this regard, as a large share of GHGs in ruminants is off-set (Escribano et al. 2022; Reyes-Palomo et al. 2022; Bronts et al. 2021). Moreover, organic farms usually promote greater use of natural resources and limit the use of chemically synthesized ingredients (antibiotics, pesticides, hormones and synthetic chemicals) (Qin et al. 2021) and promote ecosystem services such as impacts on biodiversity, soil health, pest incidence, etc.

In recent years, research on chemical composition and fatty acid (FA) profiles of CS and OS milk and dairy products has also gained attention (Ellis et al. 2006; Butler et al. 2011; Qin et al. 2021; Linehan et al. 2024). For instance, it has been well known that the OS milk contain higher levels of whey proteins, total polyunsaturated fatty acids (PUFA), n-3 PUFA and vitamin E (α-tocopherol) (Liu et al. 2018; Manuelian et al. 2020). The volatile compounds (VOCs) (Vasta et al. 2012; Villeneuve et al. 2013; Ueda et al. 2016), mineral elements and vitamins (Mogensen et al. 2012; Rey-Crespo et al. 2013; Zwierzchowski and Ametaj 2018), and FA profiles (Liu et al. 2020) in CS and OS milk production have also been investigated. Reports (Gomes et al. 2020) also suggest that OS milk production affects the microorganism composition of milk as well.

Although some positive effects of OS-derived foods are reported, the debates on superiority of OS milk and dairy products versus CS counterparts remain ongoing (Calabro and Vieri 2024). It has been shown (Butler et al. 2011; Massouras et al. 2018) that the OS milk had significantly higher PUFA concentration than those in CS milk. Regarding health-promotiog features of OS milk, Kesse-Guyot et al. (2017) conducted a study involving 62,000 participants found that high organic food consumers had slower BMI increases [BMI = −0.16, 95% confidence interval (CI): −0.32; −0.01] and a 31% (95% CI: 18%; 42%) reduced obesity risk than low OS food consumers. Rist et al. (2007) studied 2700 mothers and babies from the Netherlands, discovering that exclusive consumption of organic dairy products during pregnancy and early life corresponded to a 36% decrease in eczema prevalence at 2 years old. Organic diet association included enhanced ruminant fatty acid presence in maternal milk. However, a systemic review of 20 published paper demonstrated that the literature lacks strong evidence that organic foods are significantly more nutritious than conventional foods and organic foods may only reduce exposure to antimicrobilas and antibiotic-resistant bacteria (Smith-Spangler et al. 2013). Due to the globalization literature on milk composition and FA profile under CS and/or OS, meta-analysis offers great approach allowing researchers to identify tendencies, connections, contradictions, and verify concepts that remained unattainable via single investigations (Zwierzchowski and Ametaj 2018). Despite the regulation regarding certified organic milk, analytical differentiation has not been established and remains unclear with respect to conventional dairy systems. Therefore, this study aimed to compare CS vs OS dairy production using a systematic review and meta-analysis approach.

6.2 Materials and Methods

The current analytic review focus in controlled studies that evaluated the effect of production system (organic or conventional) on milk production and composition in ruminants. The search was carried out by three field experts to avoid the reviewer-bias in the follow scientific databases: PubMed, Google Scholar, PRIMO-UAEH, and CAB Abstracts. Also, other studies were identified from the list of references of articles found in the first search. The follow search engine was provided to each field expert to integer the studies database: "milk", "organic system", "conventional system", "milk composition", "fatty acid profile", "conjugated linolenic acid", "organic dairy products", "ruminant", and "fed management".

Only articles of indexed journals were included in the current literature review. To be considered the studies must met the follow inclusion and exclusion criteria according with Lean et al. (2014): (i) specify clearly the procedure used to randomly assign animals to treatment and control groups; (ii) least squares mean of the control and experimental group and variability measures (standard error or standard deviation); (iii) report information of sample size. A total of 37 published papers which were considered to analysis since fulfil the inclusion criteria describe above. The final database included the following output variables: milk yield (MY; l/d), fat content (FAT; g/100 g), protein content (PROT; g/100 g), lactose content (LACT; g/100 g), polyunsaturated FA (PUFA; mg/g MFA); monounsaturated FA (MUFA mg/g MFA), α-linolenic acid (ALA; mg/g MFA), vaccenic acid (VA, mg/g MFA), conjugated linoleic acid (CLA; mg/g MFA). To each output variable was captured the follow information: least squares mean, variability measures (standard error of mean or standard deviation) and number of experimental units for both production systems.

The number of available articles changed according with the output variables analysed: milk yield (n = 37), fat content (n = 30), protein content (n = 25), lactose content (n = 17), PUFA (n = 22), MUFA (n = 22), ALA (n = 20), vaccenic acid (n = 22), and CLA (n = 19).

6.2.1 Meta-Analytical Procedure

In the current work fixed and random effect models were fitted to estimate the effect size (ES), 95% confidence interval and statistical significance of ES to each outcome variable using the 'meta' package version 4.6–0 (Schwarzer 2017). In the R statistical software version 3.3.1 (R Core Team 2015). The ES was calculated as standardized mean difference (SMD) using the methods of Hedges's d to avoid the overestimation of overall effect size when the number of samples is small (Hedges 1981) for the fixed effects and DerSimonian and Laird (1986) for random effects models. Studies that reported outcome variables in the same unit of measure or was possible it's conversion were used to calculate the raw mean difference (MD), which permits ES interpretation under original measures units (Appuhamy et al. 2013).

To facility ES interpretation, the effect of organic management on productive performance was displayed in a forest plot showing the estimated SMD from the fixed and random models. The meta package (Schwarzer 2017) provides a forest plot with the SMD and weighted contribution to each study from fixed and random effect models. Publication bias was investigated using a funnel plot. The funnel plot shows graphically the precision of the estimated intervention effect, where smaller studies had a wider variance and the larger studies with less spread of variability. In absence of bias, the funnel plot should be approximately symmetrical.

The current systematic review analyses studies performed in different places with different methods and under different animal management; hence, the heterogeneity is inevitable. Heterogeneity of results among trials was reported using the I^2 statistic (Higgins and Green 2011). The I^2 represents the approximate proportion of total variability in point estimates that can be attributed to heterogeneity, which was calculates as follows:

$$I^2 = \frac{Q-(k-1)}{Q} \times 100$$

Where Q is the X^2 heterogeneity statistic and K is the number of trials. I^2 values of 25%, 50% and 75% represent small, moderate and high levels of heterogeneity, respectively. For output variables that showed substantial heterogeneity ($I^2 > 50\%$), mixed-effect models (meta-regression analysis) were constructed to explore the sources of heterogeneity in the "metafor" package (Viechtbauer 2010). Also, to quantify the effect of significant covariates on heterogeneity, I^2 statistic was calculated to random effect and mixed effect models. The covariates defined a priori in the current meta-regression were specie (cow, goat and sheep), feed management (extensive and intensive), experimental unit (herd, animal and milk sample) and neutral detergent fibre (NDF) and acid detergent fibre (ADF) of diets.

6.3 Results

With the aim of assessing the effect of CS and OS management on milk production and composition, an analytical review was carried out, followed by meta-analysis and exploration of heterogeneity using a meta-regression approach. The number of analyses varied according to the output variables analysed: milk yield (n = 39), fat content (n = 30), protein content (n = 25), lactose content (n = 17), PUFA (n = 22), MUFA (n = 22), ALA (n = 20), vaccenic acid (n = 22), and CLA (n = 19).

Our results reveal that animals managed under OS produce less milk (SMD -0.65; 95% CI: −1.01, −0.27; $p = 0.0007$) compared to those under CS. However, the concentration of milk fat, protein, and lactose were not influenced by the production system ($p > 0.05$). The content of PUFA (SMD +1.13; 95% CI: 0.33, 1.92; $p = 0.005$) and ALA (SMD +1.54; 95% CI: +0.67, +2.39; $p = 0\ .0004$) was higher in milk from OS, but both variables showed a higher heterogeneity ($I^2 > 93.0\%$). Also, we found that the content of CLA was higher in OS (SMD +3.06; 95% CI: +0.09, +6.04; $p = 0.04$) with a large heterogeneity ($I^2 > 94.0\%$) (Table 6.1).

As shown in Fig. 6.1, covariates such as animal species affected the relationship between OS farming and milk production. Decreased milk yield was higher in cows (SMD = −0.98) compared to sheep (SMD = −0.13) and goat (−0.22). Also, the nutritional characteristics of the animal diets impacted the milk production, whereas increase the NDF (−0.12*) and ADF (−0.22*) led to lower milk yield in OS production (Table 6.2). The increasing in PUFA levels of milk was more evident when both production systems were management under Intensive conditions (+14.9*).

Table 6.1 List of output variables and number of trials used in meta-analysis

Item	Number of trials(n)	Effect size (SMD ± IC)	P-value	I^2 RM	I^2 MM
Milk yield	39	−0.65 (−1.01, −0.27)	0.0007	95.4	0.0
Fat content	30	+0.06 (−0.27, 0.39)	0.72	86.3	67.6
Protein content	25	−0.16 (−0.55, +0.22)	0.41	90.8	78.8
Lactose content	17	−0.17 (−0.49, +0.14)	0.28	79.1	76.6
PUFA	22	1.13 (+0.33, +1.92)	0.005	94.0	91.3
MUFA	22	0.42 (−0.59, +1.42)	0.41	93.6	80.1
ALA	20	1.54 (+0.67, + 2.39)	0.0004	93.1	52.78
Vaccenic	21	−0.02 (−0.56, +0.53)	0.95	93.6	83.1
CLA	19	3.06 (+0.09, +6.04)	0.04	94.0	94.6

RM random model, *MM* mixed model

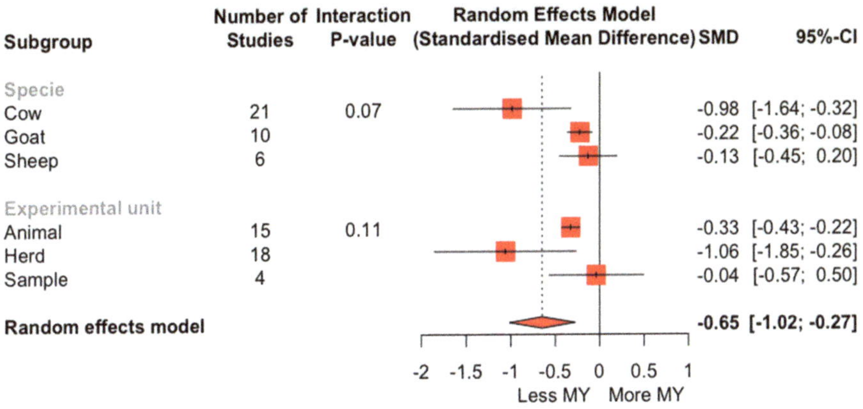

Fig. 6.1 Forest plot of milk yield (MY) from studies that compare organic and conventional management. Point size reflects the relative weighing of the study to the overall effect size estimated where larger point size represents greater weight and the combined effect size estimated including its confidence interval

Table 6.2 Meta-regression slopes and significance of covariables to outcomes with a moderate or higher heterogeneity ($I^2 > 25\%$)

Item	Specie	Management	Experimental unit	NDF	ADF
Milk yield	Goat +0.79* Sheep +0.85*		Herd −0.81*	−0.02*	−0.04*
Fat content	Cow −7.17*** Goat −2.01*	Extensive +2.04*	Sample −4.76**		
Protein content			Sample + 0.79*		
Lactose content		Intensive −0.54*			
PUFA		Intensive +14.9*	Herd +58.47* Sample −8.93*	−2.34*	
MUFA	Goat +24.76*** Sheep +14.21***		Herd +74.51***	−2.85***	
ALA	Goat −34.55***	Extensive −5.27***	Herd −25.01*** Animal −15.63***	−0.85***	
Vaccenic			Herd +64.19** Animal +42.28*	−2.51*	
CLA	Cow +14.61***	Extensive +5.46***		−0.41***	

*p<0.05, **p<0.01, ***p<0.001

Reductions in PUFA was also linked to increased dietary NDF (−2.34*). Regarding ALA content, the effect size of OS management was lower in extensive management (−5.27***). Additionally, increase in dietary NDF (−0.85***) was associated with lower ALA content in OS production. The impact of OS production was

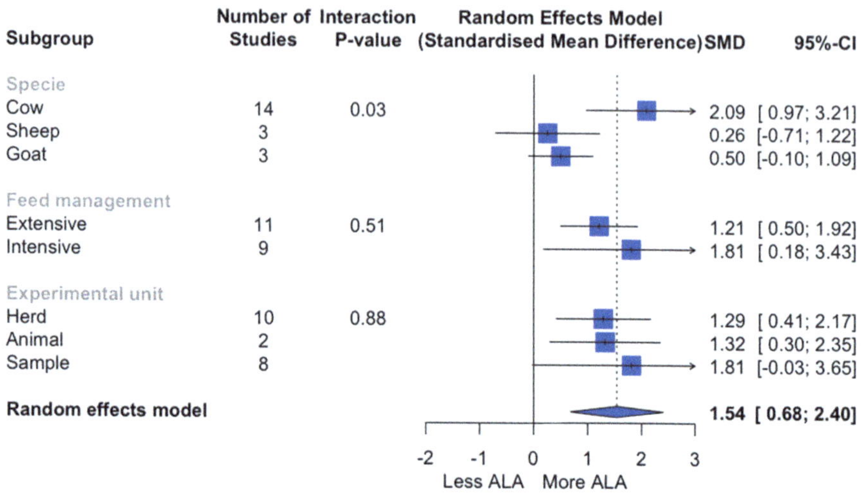

Fig. 6.2 Forest plot of α-linolenic acid (ALA) milk content from studies that compare organic and conventional management. Point size reflects the relative weighing of the study to the overall effect size estimated where larger point size represents greater weight and the combined effect size estimated including its confidence interval

generally higher when the experimental unit consisted of whole herds rather than individual animal (Table 6.2).

The effect size was higher in cows (SMD = 2.09) compared to sheep (SMD = 0.26) and goat (SMD = 0.5) for ALA concentration in milk (Fig. 6.2). Likewise, the impact of OS production on CLA content was higher in cows (SMD = 4.25) than in sheep (SMD = 1.34) and goat (SMD = 0.63) (Fig. 6.3); this difference was even more pronounced under extensive management (+5.46***).

6.4 Discussion

For dairy products to exhibit superior nutritional quality, they should contain higher amounts of some valuable nutrients and/or lower amounts of hazardous compounds. Indeed, nutrient claims can be justified if substantial quantities are proven; however, in most cases valuable milk nutrients remains largely unrecognized (Palupi et al. 2012). Milk and dairy products possessing premium nutritional quality meet the following criteria: high omega-3 to omega-6 ratios, abundant CLA and VA content, and high quantities of protein, vitamins (such as A, D, E, C, B1, B2, B11, and B12), calcium, zinc, iron, potassium, and essential oils (Haug et al. 2007; Palupi et al. 2012; Chauhan et al. 2021). The chemical composition of milk is influenced by many factors, such as the animal's diet, genetics, stage of lactation, management and season, and the interactions between them, all of which can affect the composition of the milk either in the individual animal or in the animal's environment (Schwendel et al. 2015; Srednicka-Tober et al. 2016; Brodziak et al. 2021).

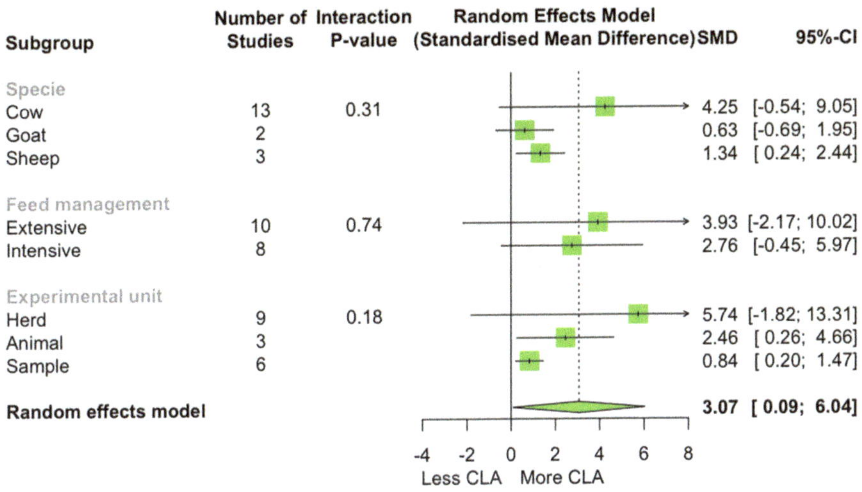

Fig. 6.3 Forest plot of conjugated linoleic acid (CLA) milk content from studies that compare organic and conventional management. Point size reflects the relative weighing of the study to the overall effect size estimated where larger point size represents greater weight and the combined effect size estimated including its confidence interval

Although, this meta-analysis could not cover all the above points, a systematic review bridges the update concept and the results available from primary studies. Hence, our optimal approach involves tailoring the previously defined dairy product nutritional quality standards to align with the insights obtained from chosen studies, specifically, below we discuss several key factors, namely fat, protein, MUFA, PUFA, ALA, CLA, and VA.

6.4.1 *Milk Yield*

Milk yield emerges as a critical determinant of the economic viability of dairy farms (Brodziak et al. 2021). Despite the existence of specialized OS dairy farms, those embrace grassland-based and boasting cows capable of yielding more than 9000 kg of milk per year (Padel et al. 2010), milk production from organically reared cows is lower, on average, than that from conventional cows (Padel et al. 2010; Linehan et al. 2024). Similarly, (Srednicka-Tober et al. 2016) using a systematic review and meta-analysis showed that the average milk yield (kg milk/cow/d or) was significantly lower in OS (−23; 95% CI −31, −15%) compared CS. Likewise, a study comparing milk yield and composition in CS and OS managed cows showed that the average milk yield was significantly lower OS production than SC (Qin et al. 2021). Also, lower milk yield in OS system has been reported several times in previous studies (Stergiadis et al. 2012; Stergiadis et al. 2015; Teter et al. 2020; Brodziak et al. 2021).

Gruber et al. (2001) demonstrated that milk yields per cow and year were identical for both herds, but milk production per area grazed was reduced in the organic herd because of lower DM yields from organic pasture and, therefore, lower stocking rates per hectare. Overall, decreased production under organic management can be traced to lower energy intake, through either less concentrate feeding or lower energy content in forages from organic systems (Stergiadis et al. 2015; Brodziak et al. 2021; Qin et al. 2021). Similarly, Angeles-Hernandez et al. (2020) found a significant reduction in milk yield in sheep fed forage-based rations, which may be related to the fact that these rations are characterized by higher fiber (neutral detergent fiber and acid detergent fiber) and lower non-structural carbohydrate (sugars and starch) content than concentrate-based rations. Therefore, lower profitability of OS milk because of lower milk yield could pose obstacles for trajectory of the global OS dairy industry (Linehan et al. 2024).

6.4.2 Fat

The fat content of milk ranges from 3.4% to 5.1%, with the FA including diacylglycerides, monoacylglycerides, phospholipids, free fatty acids, sterols, and esters (Mills et al. 2011). The predominant fatty acids in milk are saturated fatty acids (SFA), followed by unsaturated fatty acids, including monounsaturated fatty acids (MUFA) and polyunsaturated fatty acids (PUFA) (Linehan et al. 2024). It has been well established that the SFA in milk and dairy products are associated with negative consumer perceptions (Vargas-Bello-Pérez et al. 2020). Nevertheless, it is important to note that dairy fat cannot be generally described as beneficial or detrimental to human health, as the biological function of each FA should be considered separately (Schwendel et al. 2015). The total fat content of OS and CS produced milk is typically reported to not differ significantly (Müller and Sauerwein 2010; Qin et al. 2021; Linehan et al. 2024), however, there some reports showed an increased (Butler et al. 2011) or a decreased fat content in OS milk from cows (Kuczyńska et al. 2012) or small ruminants (Tudisco et al. 2010; Malissiova et al. 2015). These conflicting reports may be attributed to variation in diets, animal breeds, or the interplay of multiple factors that influence milk fat composition (Tzamaloukas et al. 2021). In confirmation, it has been well known that the concentrations of milk fat are influenced by several factors, including the animal species, the stage of lactation, genetic, and diet (Qin et al. 2021). The overall composition and quantity of FAs in milk are primarily governed by the feeding system (Palupi et al. 2012). Overall, the present results suggest that the milk fat content was not affected by the OS farming practices.

6.4.3 Protein

The proteins in milk have a wide range of biological activities such as a polyvalent role in immune function, antimicrobial, transport and adsorption of nutrients, acting as growth factors, enzymes, hormones, and antibodies (Haug et al. 2007). In the present study, the protein content of milk remained unaffected by farming practices (i.e. OS and CS). In confirmation, previous reports showed that the total protein organically and conventionally produced milk does not differ significantly (Brodziak et al. 2018; Qin et al. 2021; Linehan et al. 2024). However, Tzamaloukas et al. (2021) reported that the protein content of milk affected by farming practices due to higher grazing/forage consumption showing significantly decreased values for organic bovine and caprine organic milk. Schwendel et al. (2015) conducted a comprehensive review of various studies comparing organic and non-organic dairy cattle farms, reporting that the CS milk had a higher protein concentration. These reported increases may be partially due to the higher use of starch-based concentrate in CS diets, thereby elevating both energy intake and the protein content of milk (Linehan et al. 2024). In brief, in our study we found no change in the protein content of cattle managed under OS and/or CS. Although protein content of milk affected by genetic, breed, and lactation stage, it has been reported to be unresponsive to dietary variations and management (Walker et al. 2004).

6.4.4 Lactose

The main carbohydrate in milk is lactose, which is a reducing disaccharide composed of glucose and galactose linked by an alpha-1-4 glycosidic linkage (Szilagyi 2002). Its concentration varies from 0 to 10% w/w, but is approximately 4.6 g per 100 ml and milk and this is the only known significant source of lactose (Fox 2009). Szilagyi (2002) reported potential beneficial effects associated with dairy/lactose consumption such as possible protection against malaria, reduced osteoporosis, reduced Mental depression in women, and protection against gastric ulcers.

Generally, it has been known that lactose did not significantly alter by feeding system (Linehan et al. 2024). However, Luukkonen et al. (2005) demonstrated that Finnish organic milk contained significantly more lactose (4.79% vs. 4.74%, $p < 0.01$). Our data indicated the lack of farming systems on milk lactose levels. Similarly, Jenkins and McGuire (2006) have indicated that cow nutrition does not typically result in alterations in lactose concentration, barring exceptional situations. Supporting this claim, some studies (Nauta et al. 2006; Bilik and Łopuszańska-Rusek 2010) have reported no significant variance in lactose content between OS and CS produced milk.

6.4.5 Polyunsaturated Fatty Acids

The polyunsaturated fatty acids (PUFA) are molecules containing two or more double bonds in the hydrocarbon chain (Kapoor et al. 2021). The main PUFA are the omega-6 (ω-6) and omega-3 (ω-3) FA (Ellis et al. 2006). The difference between them is the position of the first double bond and in the fatty acid chains (Kapoor et al. 2021). The PUFA can be produced by more than one metabolic pathway, in fact some organisms possess a Δ9 desaturase that introduces a cis double bond into a saturated fatty acid (Kapoor et al. 2021). The main ω-3 FA present in milk is the α-linolenic acid (ALA; 18:3 ω-3), with smaller amounts of docosahexaenoic acid (DHA; 22:6 ω-3) and eicosapentaenoic acid (EPA; 20:5 ω-3) (Ellis et al. 2006). In human nutrition, efforts are being made to increase the consumption of ω-3 components and raise ratio of ω-3/ω-6 FA intake with the target ratio of 1:1 or 2:1 (Chilliard et al. 2007).

Diet is on the key driver of the milk FA profile; however, endogenous synthesis and the effects of physiological factors have been also reported (Chilliard et al. 2007; Benbrook et al. 2018). Comparing the FA profile of OS and CS milk production has shown significant improvement FA of OS-derived milk (Capuano et al. 2015; Srednicka-Tober et al. 2016; Puppel et al. 2017; Benbrook et al. 2018; Brodziak et al. 2021). Our results indicate an increase in PUFA with OS produced milk. In line with this, previous reports (Benbrook et al. 2018; Srednicka-Tober et al. 2016) showed that OS milk have higher PUFA, including omega-6 and omega-3, than CS milk. Kuczyńska et al. (2012) reported a lower ω-6 to ω-3 ratio (by more than 50%) and a lower level of n-6 polyunsaturated fatty acids in raw organic milk, while the ω −3 PUFA level was higher. The higher content of ω-3 PUFA in milk from OS farms that associated with a favorable of ω6:ω-3 ratio have been also confirmed (Ellis et al. 2006; Butler et al. 2011; Adler et al. 2013; Benbrook et al. 2018). Similarly, a recent paper (Tzamaloukas et al. 2021) showed that OS dairy farming positively impacted the PUFA in milk of cow, goat, and sheep. These favorable FA profile in OS milk, is mainly due to higher use of fresh pasture vegetation, including oilseed plants, such as flax (Brodziak et al. 2021; Ellis et al. 2006). The ω −6 and ω −3 ratio in milk refers to the concentrations of linoleic acid (LA) and α-linolenic acid (ALA), as these are the most abundant ω −6 and ω −3 FA (Schwendel et al. 2015). It has been also well known that the forage preservation processes (i.e. wilting, ensiling, and drying) result in significant PUFA oxidative degradation and lipolysis, driven by plant metabolism (Dewhurst et al. 2006). Consequently, fresh pasture grass, which contains higher PUFA levels, notably C18:3, than hay or silage, supports increased PUFA levels in organic dairy cattle through dietary fresh forage incorporation (Ferlay et al. 2006; Tzamaloukas et al. 2021). Overall, increasing the intake of fresh forage in dairy diets, OS managed cattle lead to elevated PUFA levels.

6.4.6 *Monounsaturated Fatty Acids*

The monounsaturated fatty acids (MUFA) are the molecules that lack a pair of hydrogen atoms in the middle part, and the free bonds are connected to a double bond between carbon atoms (Shramko et al. 2020). The American Heart Association guidelines set a limit of 20% MUFA. One reason for specific MUFA recommendations may be their potential benefit in cardio protection, as several epidemiological studies have reported that higher intakes of MUFA are associated with a lower risk of coronary heart disease (Haug et al. 2007; Mills et al. 2011; Shramko et al. 2020). In the present study, MUFA remained unaffected by farming practices (i.e. OS and/or CS) which in in agreement previously published literature (Butler et al. 2011; Liu et al. 2020; Stergiadis et al. 2015). However, some reports showed increased (Butler et al. 2019; Tudisco et al. 2010; Tsiplakou et al. 2010) or decrease (Ellis et al. 2006; Collomb et al. 2008; Adler et al. 2013) in that MUFA concentration of organically produced milk. Likewise, meta-analytical studies demonstrated contradictory results regarding MUFA content of OS and CS milk content. Palupi et al. (2012) showed that MUFA levels decreased in OS milk, while Srednicka-Tober et al. (2016) reported no significant effect between OS and CS farms, suggesting the specificity of feed components in influencing this FA group.

6.4.7 *Alpha-Linolenic Acid*

Alpha-linolenic acid (ALA) is the most abundant polyunsaturated fatty acid and an important precursor of the bioactive omega-3 PUFAs, eicosapentaenoic and docosahexaenoic acids (Rist et al. 2007). Both precursors are converted to their long-chain metabolites through a series of desaturation and elongation steps and share common enzymes for these metabolic transformations (Haug et al. 2007). Our data showed an increased ALA level in OS-derived milk. Similarly, a meta-analysis by Palupi et al. (2012) showed that OS dairy farming feeds cattle with higher fresh forage levels than CS farming a large effect size which led higher ALA concentration in milk. Also, Shingfield et al. (2007) demonstrated that the beneficial effects of pastures on milk ALA and CLA are associated with the high content of ALA in fresh forages that partly biohydrogenated into VA in the rumen, absorbed in the gut and secreted into milk.

6.4.8 *Conjugated Linoleic Acid*

Conjugated linoleic acid (CLA) is a component of milk fat containing a conjugated double bond system. It is a complex of positional and geometrical isomers (C18: 2n-6) and is often considered to have the biological benefits (Rist et al. 2007). The predominant source of CLA in the human diet is derived from ruminant foods, accounting for approximately 75% of the total (Haug et al. 2007). Adler et al. (2013) found differences in C18:0, C18:1 cis-9, ALA, EPA, and DHA content of OS and CS milk. The concentrations of C18:0 and C18:1 cis-9 were increased in CS-derived milk. On the other hand, OS-derived milk showed higher concentrations of ALA, EPA and DHA. Our results were in line with Srednicka-Tober et al. (2016), whose also found significantly higher total CLA in OS milk. Likewise, several studies confirm that the OS milk contains higher CLA content than CS milk (Palupi et al. 2012; Tzamaloukas et al. 2021; Linehan et al. 2024). Pasture grazing was also showed to enhance bovine milk CLA content in low-input mountain farms of Poland (Rutkowska et al. 2012). Overall, there is now a sufficiently large body of evidence confirm the favourable imact of OS farming on CLA content of milk (Palupi et al. 2012; Srednicka-Tober et al. 2016).

6.4.9 Vaccenic

The main isomer in milk fat is vaccenic acid, (18:1, 11t, VA), but trans double bonds in position 4–16 are also observed in low concentrations in milk fat (Rist et al. 2007). Evidence shows that after absorption, vaccenic acid can be converted into cis-9, trans11-CLA by enzyme 9 desaturase (Srednicka-Tober et al. 2016). Vaccenic acid has a dual role in metabolism as it is both a precursor for 9c,11 t-CLA and a trans fatty acid (Rist et al. 2007). Our results showed no significant size effect of OS and/or CS on VA concentration in milk. In contrast, Palupi et al. (2012) reported that the OS dairy product contains significantly higher VA (0.51 ± 0.16) compared to the CS product. Another meta-analysis research also showed significantly higher total VA in organic milk (Srednicka-Tober et al. 2016). However, in line with our results, Martini et al. (2023) reported that the VA concentration did not significantly differ in CS and OS produced milk.

6.4.10 Sources of Heterogeneity

It should be noted that comparing OS and CS dairy farming is difficult due to many accompanying factors (Brodziak et al. 2021). Our data suggests that the variation in milk yield and composition following different farming practices (i.e. OS and/or SC) were mainly due to the animal species and feeding management. As such, the effect size was higher in cows compare to sheep and goat for yield, CLA, and ALA concentration in milk. Our meta-regression results also show that the dietary characteristics affects milk yield in OS production, as increase in dietary NDF and ADF led to lower milk yield. Organic dairy farming provides cattle with significantly higher levels of fresh forage compared to conventional farming practices, as evidenced by a substantial effect size. This suggests that, despite the variability in regulation and implementation of organic dairy farming, there is a consistent trend towards enhanced forage intake within the organic dairy sector (Palupi et al. 2012). It has been also reported that difference between OS and CS dairy production systems is partly associated with feeding practices (Schwendel et al. 2015; Brodziak et al. 2021; Martini et al. 2023).

Confirming our findings, research has demonstrated that OS dairy production is associated with an increased intake PUFA (Dewhurst et al. 2006). This higher PUFA consumption is attributed to the elevated levels of unsaturated fatty acids, particularly ALA and other omega-3 fatty acids, present in fresh forage. Forage, especially young grass, contains significant amounts of PUFA and higher bioavailability of omega-3 fatty acids (Haug et al. 2007). The high linolenic acid (cis-9 cis-12 cis-15 C18:3) content of young grasses promotes CLA production and its trans-11 18:1 precursor. The linoleic and linolenic acids in forage are hydrogenated in the rumen and produce vaccenic acid, which is used as a precursor in the synthesis of CLA in adipose tissue and mammary glands due to the Δ9 desaturase, which has greater activity in the mammary gland in lactating animals (Angeles-Hernandez et al. 2020).

Therefore, the contents of the most of milk compounds in dairy production systems are highly dependent of the feeding management, mainly the associated with the FA profile. It should be recognized that discrepancies observed in meta-analyses of milk composition parameters may be due to two sources: the limited availability of information and the inherent variability in production systems and livestock diets (Schwendel et al. 2015). This observation underscores the necessity for (1) further well-designed studies delivering substantial additional primary data sets, (2) reporting of measures of variance in publications and (3) the establishment of registers of primary research.

6.5 Conclusion

Taken together, results of this systematic review suggest that current product categories and labels, milk produced under OS farming conditions has a higher content of beneficial components to human health, mainly associated with their FA profile. Finally, the difference in animal species and feeding management between CS and OS dairy production is suspected to be the reason behind this evidence.

References

Adler SA, Jensen SK, Govasmark E, Steinshamn H (2013) Effect of short-term versus long-term grassland management and seasonal variation in organic and conventional dairy farming on the composition of bulk tank milk. J Dairy Sci 96(9):5793–5810. https://doi.org/10.3168/jds.2012-5765

Akram M, Sami M, Ahmed O, Onyekere PF, Egbuna C (2020) Health benefits of milk and milk products. In: Functional foods and nutraceuticals: bioactive components, formulations and innovations, pp 211–217

Angeles-Hernandez JC, Alberto RV, Kebreab E, Appuhamy JARN, Dougherty HC, Castelan-Ortega O, Gonzalez-Ronquillo M (2020) Effect of forage to concentrate ratio and fat supplementation on milk composition in dairy sheep: a meta-analysis. Livest Sci 238:104069. https://doi.org/10.1016/j.livsci.2020.104069

Appuhamy JRN, Strathe AB, Jayasundara S, Wagner-Riddle C, Dijkstra J, France J, Kebreab E (2013) Anti-methanogenic effects of monensin in dairy and beef cattle: a meta-analysis. J Dairy Sci 96(8):5161–5173. https://doi.org/10.3168/jds.2012-5923

Benbrook CM, Davis DR, Heins BJ, Latif MA, Leifert C, Peterman L et al (2018) Enhancing the fatty acid profile of milk through forage-based rations, with nutrition modeling of diet outcomes. Food Sci Nutr 6(3):681–700. https://doi.org/10.1002/fsn3.610

Bilik K, Łopuszańska-Rusek M (2010) Effect of organic and conventional feeding of Red-and-White cows on productivity and milk composition. Ann Anim Sci 10(4):441–458

Bórawski P, Bórawski MB, Parzonko A, Wicki L, Rokicki T, Perkowska A, Dunn JW (2021) Development of organic Milk production in Poland on the background of the EU. Agriculture 11(4):323. https://doi.org/10.3390/agriculture11040323

Brodziak A, Król J, Litwińczuk Z, Barłowska J (2018) Differences in bioactive protein and vitamin status of milk from certified organic and conventional farms. Int J Dairy Technol 71(2):321–332. https://doi.org/10.1111/1471-0307.12462

Brodziak A, Wajs J, Zuba-Ciszewska M, Król J, Stobiecka M, Jańczuk A (2021) Organic versus conventional raw cow milk as material for processing. Animals 11(10):2760. https://doi.org/10.3390/ani11102760

Bronts S, Gerbens-Leenes PW, Guzmán-Luna P (2021) The water, land and carbon footprint of conventional and organic dairy systems in The Netherlands and Spain. A case study into the consequences of ecological indicator selection and methodological choices. Energy Nexus 11:100217. https://doi.org/10.1016/j.nexus.2023.100217

Butler G, Stergiadis S, Seal C, Eyre M, Leifert C (2011) Fat composition of organic and conventional retail milk in Northeast England. J Dairy Sci 94(1):24–36. https://doi.org/10.3168/jds.2010-3331

Butler G, Stergiadis S, Chatzidimitriou E, Franceschin E, Davis HR, Leifert C, Steinshamn H (2019) Differing responses in milk composition from introducing rapeseed and naked oats to conventional and organic dairy diets. Sci Rep 9(1):8115. https://doi.org/10.1038/s41598-019-44567-8

Calabro G, Vieri S (2024) Limits and potential of organic farming towards a more sustainable European agri-food system. Br Food J 126(1):223–236. https://doi.org/10.1108/bfj-12-2022-1067

Capuano E, Gravink R, Boerrigter-Eenling R, van Ruth SM (2015) Fatty acid and triglycerides profiling of retail organic, conventional and pasture milk: implications for health and authenticity. Int Dairy J 42:58–63. https://doi.org/10.1016/j.idairyj.2014.11.002

Chauhan S, Powar P, Mehra R (2021) A review on nutritional advantages and nutraceutical properties of cow and goat milk. Int J Appl Res 7:101–105. https://doi.org/10.22271/allresearch.2021.v7.i10b.9025

Chilliard Y, Glasser F, Ferlay A, Bernard L, Rouel J, Doreau M (2007) Diet, rumen biohydrogenation and nutritional quality of cow and goat milk fat. Eur J Lipid Sci Technol 109(8):828–855. https://doi.org/10.1002/ejlt.200700080

Collomb M, Bisig W, Bütikofer U, Sieber R, Bregy M, Etter L (2008) Fatty acid composition of mountain milk from Switzerland: comparison of organic and integrated farming systems. Int Dairy J 18(10–11):976–982. https://doi.org/10.1016/j.idairyj.2008.05.010

De Boer IJ (2003) Environmental impact assessment of conventional and organic milk production. Livest Prod Sci 80(1–2):69–77. https://doi.org/10.1016/S0301-6226(02)00322-6

DerSimonian R, Laird N (1986) Meta-analysis in clinical trials. Control Clin Trials 7(3):177–188. https://doi.org/10.1016/0197-2456(86)90046-2

Dewhurst RJ, Shingfield KJ, Lee MR, Scollan ND (2006) Increasing the concentrations of beneficial polyunsaturated fatty acids in milk produced by dairy cows in high-forage systems. Anim Feed Sci Technol 131(3–4):168–206. https://doi.org/10.1016/j.anifeedsci.2006.04.016

Ellis KA, Innocent G, Grove-White D, Cripps P, McLean WG, Howard CV, Mihm M (2006) Comparing the fatty acid composition of organic and conventional milk. J Dairy Sci 89(6):1938–1950. https://doi.org/10.3168/jds.s0022-0302(06)72261-5

Escribano M, Horrillo A, Mesias FJ (2022) Greenhouse gas emissions and carbon sequestration in organic dehesa livestock farms. Does technical-economic management matters? J Clean Prod 372:133779. https://doi.org/10.1016/j.jclepro.2022.133779

FAO (2024) The World of organic agriculture 2024. https://www.fao.org/family-farming/detail/en/c/1678159/

Ferlay A, Martin B, Pradel P, Coulon JB, Chilliard Y (2006) Influence of grass-based diets on milk fatty acid composition and milk lipolytic system in Tarentaise and Montbéliarde cow breeds. J Dairy Sci 89(10):4026–4041. https://doi.org/10.3168/jds.S0022-0302(06)72446-8

Fox PF (2009) Lactose: chemistry and properties. In: Advanced dairy chemistry: volume 3: lactose, water, salts and minor constituents. Springer, New York, pp 1–15

Gomes SI, van Bodegom PM, van Agtmaal M, Soudzilovskaia NA, Bestman M, Duijm E et al (2020) Microbiota in dung and milk differ between organic and conventional dairy farms. Front Microbiol 11:1746. https://doi.org/10.3389/fmicb.2020.01746

Gruber L, Steinwender R, Guggenberger T, Häusler J, Schauer A (2001) Comparison of organic and conventional farming on a grassland farm. In: 2nd communication: feed intake, milk yield, health and fertility parameters

Gutierrez RT, Vega YLS, Schettino BB, Perez JJG, Ruiz Rojas JL, Radilla Vazquez CC, Del Muro Delgado R (2022) Fatty acids and conjugated linoleic acid in organic milk produced in southeastern Mexico. Revista Mvz Cordoba 27(2). https://doi.org/10.21897/rmvz.2541

Haug A, Høstmark AT, Harstad OM (2007) Bovine milk in human nutrition–a review. Lipids Health Dis 6:1–16. https://doi.org/10.1186/1476-511x-6-25

Hedges LV (1981) Distribution theory for Glass's estimator of effect size and related estimators. J Educ Stat 6(2):107–128. https://doi.org/10.1186/1476-511X-6-25

Higgins JPT, Green S (2011) Cochrane handbook for systematic reviews of interventions version 5.1.0 [updated March 2011]. Cochrane Collaboration, London. Available from. www.handbook.cochrane.org. https://doi.org/10.1002/9780470712184

Jenkins TC, McGuire MA (2006) Major advances in nutrition: impact on milk composition. J Dairy Sci 89(4):1302–1310. https://doi.org/10.3168/jds.S0022-0302(06)72198-1

Kapoor B, Kapoor D, Gautam S, Singh R, Bhardwaj S (2021) Dietary polyunsaturated fatty acids (PUFAs): uses and potential health benefits. Curr Nutr Rep 10:232–242. https://doi.org/10.1007/s13668-021-00363-3

Kesse-Guyot E, Baudry J, Assmann KE, Galan P, Hercberg S, Lairon D (2017) Prospective association between consumption frequency of organic food and body weight change, risk of overweight or obesity: results from the NutriNet-Santé Study. Br J Nutr 117(2):325–334. https://doi.org/10.1017/S0007114517000058

Kuczyńska B, Puppel K, Gołębiewski M, Metera E, Sakowski T, Słoniewski K (2012) Differences in whey protein content between cow's milk collected in late pasture and early indoor feeding season from conventional and organic farms in Poland. J Sci Food Agric 92(14):2899–2904. https://doi.org/10.1002/jsfa.5663

Lambotte M, De Cara S, Brocas C, Bellassen V (2023) Organic farming offers promising mitigation potential in dairy systems without compromising economic performances. J Environ Manag 334:117405. https://doi.org/10.1016/j.jenvman.2023.117405

Lean IJ, Thompson JM, Dunshea FR (2014) A meta-analysis of zilpaterol and ractopamine effects on feedlot performance, carcass traits and shear strength of meat in cattle. PLoS One 9(12):e115904. https://doi.org/10.1371/journal.pone.0115904

Linehan K, Patangia DV, Ross RP, Stanton C (2024) Production, composition and nutritional properties of organic milk: a critical review. Food Secur 13(4):550. https://doi.org/10.3390/foods13040550

Liu N, Parra HA, Pustjens A, Hettinga K, Mongondry P, Van Ruth SM (2018) Evaluation of portable near-infrared spectroscopy for organic milk authentication. Talanta 184:128–135. https://doi.org/10.1016/j.talanta.2018.02.097

Liu N, Pustjens AM, Erasmus SW, Yang Y, Hettinga K, van Ruth SM (2020) Dairy farming system markers: the correlation of forage and milk fatty acid profiles from organic, pasture and conventional systems in The Netherlands. Food Chem 314:126153. https://doi.org/10.1016/j.foodchem.2019.126153

Luukkonen J, Kemppinen A, Kärki M, Laitinen H, Mäki M, Sivelä S et al (2005) The effect of a protective culture and exclusion of nitrate on the survival of enterohemorrhagic E. coli and Listeria in Edam cheese made from Finnish organic milk. Int Dairy J 15(5):449–457. https://doi.org/10.1016/j.idairyj.2004.09.005

Malissiova E, Tzora A, Katsioulis A, Hatzinikou M, Tsakalof A, Arvanitoyannis IS et al (2015) Relationship between production conditions and milk gross composition in ewe's and goat's organic and conventional farms in Central Greece. Dairy Sci Technol 95:437–450. https://doi.org/10.1007/s13594-015-0224-7

Manuelian CL, Penasa M, da Costa L, Burbi S, Righi F, De Marchi M (2020) Organic livestock production: a bibliometric review. Animals 10(4):618. https://doi.org/10.3390/ani10040618

Martini M, Altomonte I, Sodi I, Vasylieva Y, Salari F (2023) Sterol, tocopherol, and bioactive fatty acid differences between conventional, high-quality, and organic cow milk. J Dairy Sci 106(12):8239–8248. https://doi.org/10.3168/jds.2023-23378

Massouras T, Maragoudakis S, Hadjigeorgiou I (2018) Differences in sheep milk characteristics focusing on fatty acid profile between conventional and organic farming system. Arch Dairy Res Technol. https://doi.org/10.29011/ADRT-104

Mie A, Andersen HR, Gunnarsson S, Kahl J, Kesse-Guyot E, Rembiałkowska E et al (2017) Human health implications of organic food and organic agriculture: a comprehensive review. Environ Health 16:1–22. https://doi.org/10.1186/s12940-017-0315-4

Mills S, Ross RP, Hill C, Fitzgerald GF, Stanton C (2011) Milk intelligence: mining milk for bioactive substances associated with human health. Int Dairy J 21(6):377–401. https://doi.org/10.1016/j.idairyj.2010.12.011

Mogensen L, Kristensen T, Søegaard K, Jensen SK, Sehested J (2012) Alfa-tocopherol and beta-carotene in roughages and milk in organic dairy herds. Livest Sci 145(1–3):44–54. https://doi.org/10.1016/j.livsci.2011.12.021

Müller U, Sauerwein H (2010) A comparison of somatic cell count between organic and conventional dairy cow herds in West Germany stressing dry period related changes. Livest Sci 127(1):30–37. https://doi.org/10.1016/j.livsci.2009.08.003

Nauta WJ, Veerkamp RF, Brascamp EW, Bovenhuis H (2006) Genotype by environment interaction for milk production traits between organic and conventional dairy cattle production in The Netherlands. J Dairy Sci 89(7):2729–2737. https://doi.org/10.3168/jds.S0022-0302(06)72349-9

Padel S, Vine J, Huber B, Stolze M, Jespersen LM, Rüegg E, …, Belliere SR (2010) The European regulatory framework and its implementation in influencing organic inspection and certification systems in the EU. http://orgprints.org/18359/1/padel-etal-2010-certcost.pdf

Palupi E, Jayanegara A, Ploeger A, Kahl J (2012) Comparison of nutritional quality between conventional and organic dairy products: a meta-analysis. J Sci Food Agric 92(14):2774–2781. https://doi.org/10.1002/jsfa.5639

Puppel K, Sakowski T, Kuczyńska B, Grodkowski G, Gołębiewski M, Barszczewski J et al (2017) Degrees of antioxidant protection: a 2-year study of the bioactive properties of organic milk in Poland. J Food Sci 82(2):523–528. https://doi.org/10.1111/1750-3841.13594

Qin N, Faludi G, Beauclercq S, Pitt J, Desnica N, Pétursdóttir Á et al (2021) Macromineral and trace element concentrations and their seasonal variation in milk from organic and conventional dairy herds. Food Chem 359:129865. https://doi.org/10.1016/j.foodchem.2021.129865

R Core Team (2015) R: A language and environment for statistical computing [Internet]. R Foundation for Statistical Computing, Vienna. http://www.r-project.org

Rey-Crespo F, Miranda M, López-Alonso M (2013) Essential trace and toxic element concentrations in organic and conventional milk in NW Spain. Food Chem Toxicol 55:513–518. https://doi.org/10.1016/j.fct.2013.01.040

Reyes-Palomo C, Aguilera E, Llorente M, Díaz-Gaona C, Moreno G, Rodríguez-Estévez V (2022) Carbon sequestration offsets a large share of GHG emissions in dehesa cattle production. J Clean Prod 358:131918. https://doi.org/10.1016/j.jclepro.2022.131918

Rist L, Mueller A, Barthel C, Snijders B, Jansen M, Simoes-Wüst AP et al (2007) Influence of organic diet on the amount of conjugated linoleic acids in breast milk of lactating women in The Netherlands. Br J Nutr 97(4):735–743. https://doi.org/10.1017/S0007114507433074

Rutkowska J, Adamska A, Bialek M (2012) Fatty acid profile of the milk of cows reared in the mountain region of Poland. J Dairy Res 79(4):469–476. https://doi.org/10.1017/S0022029912000428

Schwarzer G (2017) Meta: general package for meta-analysis. https://cran.r-project.org/web/packages/meta/index.html

Schwendel BH, Wester TJ, Morel PCH, Tavendale MH, Deadman C, Shadbolt NM, Otter DE (2015) Invited review: organic and conventionally produced milk—an evaluation of factors influencing milk composition. J Dairy Sci 98(2):721–746. https://doi.org/10.3168/jds.2014-8389

Shingfield KJ, Ahvenjärvi S, Toivonen V, Vanhatalo A, Huhtanen P (2007) Transfer of absorbed cis-9, trans-11 conjugated linoleic acid into Milk is biologically more efficient than endogenous synthesis from absorbed Vaccenic acid in lactating cows, 3. J Nutr 137(5):1154–1160. https://doi.org/10.1093/jn/137.5.1154

Shramko VS, Polonskaya YV, Kashtanova EV, Stakhneva EM, Ragino YI (2020) The short overview on the relevance of fatty acids for human cardiovascular disorders. Biomol Ther 10(8):1127. https://doi.org/10.3390/biom10081127

Smith-Spangler C, Brandeau ML, Hunter JE, Bavinger JC, Pearson M, Eschbach PJ, Sundaram V, Liu H, Schirmer P, Stave C (2013) Are organic foods safer or healthier than conventional. Ann Intern Med 157(157):144–145. https://doi.org/10.7326/0003-4819-157-5-201209040-00007

Srednicka-Tober D, Barański M, Seal CJ, Sanderson R, Benbrook C, Steinshamn H, Gromadzka-Ostrowska J, Rembiałkowska E, Skwarło-Sońta K, Eyre M (2016) Higher PUFA and n-3 PUFA conjugated linoleic acid, tocopherol and iron, but lower iodine and selenium concentrations in organic milk: a systematic literature review and meta- and redundancy analyses. Br J Nutr 115:1043–1060. https://doi.org/10.1017/S0007114516000349

Stergiadis S, Leifert C, Seal CJ, Eyre MD, Nielsen JH, Larsen MK et al (2012) Effect of feeding intensity and milking system on nutritionally relevant milk components in dairy farming systems in the north east of England. J Agric Food Chem 60(29):7270–7281. https://doi.org/10.1021/jf301053b

Stergiadis S, Leifert C, Seal CJ, Eyre MD, Larsen MK, Slots T et al (2015) A 2-year study on milk quality from three pasture-based dairy systems of contrasting production intensities in Wales. J Agric Sci 153(4):708–731. https://doi.org/10.1017/S0021859614000963

Szilagyi A (2002) Lactose—a potential prebiotic. Aliment Pharmacol Ther 16(9):1591–1602. https://doi.org/10.1046/j.1365-2036.2002.01321.x

Teter A, Barłowska J, Król J, Brodziak A, Rutkowska J, Litwińczuk Z (2020) Volatile compounds and amino acid profile of short-ripened farmhouse cheese manufactured from the milk of the White-Backed native cow breed. LWT 129:109602. https://doi.org/10.1016/j.lwt.2020.109602

Tsiplakou E, Kotrotsios V, Hadjigeorgiou I, Zervas G (2010) Differences in sheep and goats milk fatty acid profile between conventional and organic farming systems. J Dairy Res 77(3):343–349. https://doi.org/10.1017/S0022029910000270

Tudisco R, Cutrignelli MI, Calabrò S, Piccolo G, Bovera F, Guglielmelli A et al (2010) Influence of organic systems on milk fatty acid profile and CLA in goats. Small Rumin Res 88(2–3):151–155. https://doi.org/10.1016/j.smallrumres.2009.12.023

Tzamaloukas O, Neofytou MC, Simitzis PE, Miltiadou D (2021) Effect of farming system (organic vs. conventional) and season on composition and fatty acid profile of bovine, caprine and ovine milk and retail Halloumi cheese produced in Cyprus. Food Secur 10(5):1016. https://doi.org/10.3390/foods10051016

Ueda Y, Asakuma S, Miyaji M, Akiyama F (2016) Effect of time at pasture and herbage intake on profile of volatile organic compounds of dairy cow milk. Anim Sci J 87(1):117–125. https://doi.org/10.1111/asj.12396

Vargas-Bello-Pérez E, Faber I, Osorio JS, Stergiadis S (2020) Consumer knowledge and perceptions of milk fat in Denmark, the United Kingdom, and the United States. J Dairy Sci 103(5):4151–4163. https://doi.org/10.3168/jds.2019-17549

Vasta V, D'Alessandro AG, Priolo A, Petrotos K, Martemucci G (2012) Volatile compound profile of ewe's milk and meat of their suckling lambs in relation to pasture vs. indoor feeding system. Small Rumin Res 105(1–3):16–21. https://doi.org/10.1016/j.smallrumres.2012.02.010

Viechtbauer W (2010) Conducting meta-analyses in R with the metafor package. J Stat Softw 36(3):1–48. https://doi.org/10.18637/jss.v036.i03

Villeneuve MP, Lebeuf Y, Gervais R, Tremblay GF, Vuillemard JC, Fortin J, Chouinard PY (2013) Milk volatile organic compounds and fatty acid profile in cows fed timothy as hay, pasture, or silage. J Dairy Sci 96(11):7181–7194. https://doi.org/10.3168/jds.2013-6785

Walker GP, Dunshea FR, Doyle PT (2004) Effects of nutrition and management on the production and composition of milk fat and protein: a review. Aust J Agric Res 55(10):1009–1028. https://doi.org/10.1071/AR03173

Zwierzchowski G, Ametaj BN (2018) Minerals and heavy metals in the whole raw milk of dairy cows from different management systems and countries of origin: a meta-analytical study. J Agric Food Chem 66(26):6877–6888. https://doi.org/10.1021/acs.jafc.8b00904

Chapter 7
Organic Dairy Production in Europe. The Challenge of Conversion from Conventional Systems

Ruth Rodríguez-Bermúdez, Marta Miranda, Marta López-Alonso, Ramiro Fouz, and Inmaculada Orjales

Abstract The number of organic dairy farms in Europe has more than doubled in the last decade, due to the economic instability of conventional dairy farming, which is highly dependent on international input costs. Organic farming that includes livestock grazing provides an opportunity for many small farmers to compete with large intensive farms in a separate milk payment system.

Many conventional dairy producers have converted to organic systems in the last decade. Some have recently converted or are undergoing the transition and are applying organic regulations to existing facilities and livestock. These producers will have to cope with many challenges in order to increase the profitability of their systems. Organic production is a pasture-based system in which the use of chemical products and conventional treatments is either strongly restricted or prohibited. Organic dairy farmers must aim to prevent (rather than treat) diseases in their livestock and adapt to using a mainly forage-based diet. Using cow breeds that are better adapted to organic conditions therefore appears essential.

The objective of this review is to provide recommendations for the conversion from conventional to organic dairy systems in Europe, in regard to disease prevention, grazing management systems and breed adaptation.

Keywords Conversion challenges · Grazing systems · Organic dairy production · Reproductive efficiency · Sustainability · Udder health

---

R. Rodríguez-Bermúdez · M. Miranda (✉) · R. Fouz · I. Orjales
Department of Anatomy, Animal Production and Clinical Veterinary Sciences, Lugo, Spain
e-mail: marta.miranda@usc.es

M. López-Alonso
Department of Animal Pathology. Faculty of Veterinary, University of Santiago de Compostela Campus Terra, Lugo, Spain

C. Palacios Riocerezo, M. Gonzalez Ronquillo (eds.), *Traditional Livestock Production*, https://doi.org/10.1007/978-3-031-88275-3_7

7.1 Introduction

Organic production is a farm management and food production system that combines best environmental practices to protect environmental resources. In terms of livestock breeding, it attaches greater importance to animal welfare standards and biodiversity preservation than in conventional systems. Organic farming promotes the use of local feed resources, prohibits the use of pesticides and greatly limits the use of chemical products such as fertilizers and allopathic treatments. However, although nutrition and animal health are regulated by the European Union (EU), in regard to breed choice, the use of indigenous breeds is merely a recommendation (European Commission 2018, 2021).

Organic agriculture has undergone remarkable growth in the last decade: between 2012 and 2021, the number of hectares of land dedicated to organic agriculture increased by 68.3%, and it was estimated that organic cattle production increased by 6% in the same period (EUROSTAT 2023). The European Commission indicates that it may be easier and cheaper to convert extensive grass-fed systems to organic systems (European Commission 2023), identifying the opportunity of conversion from pasture-based conventional systems to organic production.

In May 2020, the European Commission presented the European Green Deal designed to ensure carbon neutrality in agriculture and to render the EU food system more sustainable across the value chain. While indicating food security and safety as priorities, the strategy's main goal is to render European agriculture more sustainable by applying the goals of organic agriculture, reducing the use of pesticides, fertilizers and antimicrobials by 50% and increasing the share of organic agriculture to up to 25% by 2030 (Aerni 2023). Encouraging conversion of conventional to organic agriculture systems is a strategy used by the EU to face up to climate change and to ensure the availability of safe food products for the population.

In recent years, the shortage of raw materials, such as cereals and fuel, has placed dairy farmers in a critical situation (European Council 2023). Some dairy producers affected by all of the recent setbacks view organic production as an opportunity to reduce spending on concentrates and fuel by focusing on using their own farm resources, while also receiving higher profits and subsidies by selling organic milk. In contrast to conventional agriculture, organic agriculture has a significant, largely untapped potential to address multiple crisis facing our society, including climate change, health and economic struggles (Brook and Johnson 2022).

Considering the challenges that conventional dairy farmers must face regarding conversion of their farms to organic dairy systems, the objective of this chapter is to analyse the most important points of conversion from a veterinary perspective, taking into consideration EU regulations, animal health, nutrition and breed choice.

7.2 European Union Regulations

The main aspects of organic regulation in the EU are considered in Regulation (EU) 2018/848 on organic production and labelling of organic products (European Commission 2018) and in the Commission Implementing Regulation (EU) 2021/1165 of 15 July 2021 authorising certain products and substances for use in organic production and establishing lists of these (C/2021/5149) (European Commission 2021). Regulation (EU) 2018/848 (European Commission 2018) establishes the principles of organic production and lays down the rules concerning organic production, related certification and the use of indications referring to organic production in labelling and advertising. According to this regulation, organic production has several general objectives, thus contributing to protecting the environment and the climate, maintaining the long-term fertility of soils, protecting biodiversity, contributing to high animal welfare standards and encouraging the preservation of rare and native breeds in danger of extinction.

Organic crops should be produced in living soil (except for crops naturally grown in water) and/or fertilized with materials and products allowed in organic production, while not exceeding the established limits. Moreover, the regulations prohibit the use of mineral fertilizers (European Commission 2018, 2021) and pesticides (European Commission 2007, 2021; European Parliament 2009). The use of genetically modified organisms (GMOs), products produced from GMOs and products produced by GMOs is also prohibited in organic production (European Commission 2018).

In the case of livestock, all animals must be born (or hatched) and raised in organic production units, except during the conversion period. Reproduction must be by natural methods, although artificial insemination is allowed. Embryo transfer and cloning are not allowed. The use of hormones or other substances with similar effects is prohibited, except as a form of veterinary therapeutic treatment in case of individual animals (European Commission 2018).

Regarding the choice of breeds or strains, Regulation (EU) 2018/848 (European Commission 2018) recommends giving preference to breeds or strains with a high degree of genetic diversity and to consider the capacity of the animals to adapt to local conditions, their breeding value, longevity, vitality and resistance to disease or health problems, all without impairment of their welfare. Breeds or strains of animals should be selected to prevent specific diseases or health problems associated with some breeds or strains used in intensive production (e.g. spontaneous abortion and difficult births requiring caesarean operations). However, the preference for indigenous breeds and strains is not regulated and is merely a recommendation.

In addition to nutrition, in the case of bovine species, at least 70% of the feed should be produced on the farm itself. If this is not feasible, or such feed is not available, it should be produced in cooperation with other organic or in-conversion production units and feed operators using feed and feed material from the same region. Animals must have access to pasture for grazing whenever conditions allow. At least 60% of the dry matter (DM) in daily rations should consist of roughage, fresh or dried fodder, or silage. This percentage may be reduced to 50% for animals in dairy production for a maximum period of 3 months at the early lactation stage (European Commission 2018).

In relation to health care, disease prevention should be based on breed and strain selection, husbandry management practices, high-quality feed, exercise, appropriate stocking density and adequate and appropriate housing maintained in hygienic conditions. Chemically synthesised allopathic veterinary medicinal products, such as antibiotics and boluses of synthesised allopathic chemical molecules, should not be used as preventive treatments. In the case of failure of preventive measures, if an animal become ill, chemically synthesised allopathic veterinary medical products can be used under strict conditions and under the responsibility of a veterinarian, when the use of phytotherapeutic, homeopathic and other products is not appropriate. In a similar vein, with the exception of vaccinations, treatments for parasites and compulsory eradication schemes, when a dairy cow or a group of dairy cattle receives more than three courses of treatments with chemically synthesised allopathic veterinary medical products (including antibiotics) within 12 months, neither the livestock concerned nor produce derived from such livestock can be sold as organic products, and a withdrawal period must be applied. The withdrawal period between the last administration of a chemically synthesised allopathic veterinary medical product (including antibiotics) under normal conditions of use, and the production of organically produced foodstuffs is twice the withdrawal period referred to in Article 11 of Directive 2001/82/EC and lasts at least 48 h (European Commission 2018).

7.3 Treatment Reduction

As mentioned above in the section on European Union regulations, farmers converting from conventional to organic production system must accept restrictions and prohibitions applied to many veterinary treatments (European Commission 2018, 2021). A detailed account all pathologies that may affect dairy cattle is outside the scope of this review, and we have selected two common examples showing how farmers must adapt to organic medical restrictions and which could be used as models for other diseases.

7.3.1 Udder Health

Mastitis is the most frequent and costly disease for dairy producers, in both conventional and organic dairy systems. Although multiple strategies can be used to prevent acquisition of new intramammary infections, unfortunately antibiotic therapy remains one of the most effective tools for treating and controlling contagious mastitis on conventional dairy farms, and the classic therapy for this disease involves administration of large amounts of antibiotics (Roesch et al. 2007; Orjales et al. 2016; Ruegg 2017). However, after conversion to organic farming, producers must adapt to minimal use of antibiotics to comply with EU regulations. Due to restrictions in antibiotic therapies and prohibition of blanket dry-cow therapy in organic farming, preventive management practices are essential in order to maintain udder health (Cicconi-Hogan et al. 2013).

The main parameters used to indicate udder health status are somatic cell counts (SCCs) and the associated linear score (LS). High SCC and/or LS values are indicative of bacterial infection (Ali and Shook 1980; Caraviello et al. 2005). Although these udder health indicators have been compared in organic and conventional production systems worldwide (Bennedsgaard et al. 2010; Villar and López-Alonso 2015; Orjales et al. 2016, 2017), no general consensus has not been reached on this topic. Thus, although some researchers have reported lower SCCs in organic conditions (Hamilton et al. 2006) or scarce differences between both production systems (Vaarst et al. 2006), most have reported the opposite patterns (Roesch et al. 2007; Rozzi et al. 2007; Orjales et al. 2016). Specifically, SCCs were found to be higher in organic than in conventional farm systems, which may indicate a higher incidence of mastitis (Villar and López-Alonso 2015; Villar et al. 2016).

Orjales et al. (2016) found that SCCs depend on the type of production system and are also strongly influenced by the treatments used on the farms, primarily antibiotics. In fact, SCCs tend to be lower on farms that use antibiotic treatments (both organic and conventional farms) than on farms that do not use antibiotics. In this respect, the difference between use or non use of antibiotics increases with the number of lactations. In the case of heifers, SCCs are higher in conventional systems than in organic systems in which antibiotics are used or not, used; however, in multiparous cows with more than 3 lactations, those in organic systems that did not receive antibiotics had higher SCCs (14.3%) than both organically reared animals administered antibiotics (7.1%) and conventionally bred animals (4.5%) (Orjales et al. 2016).

The above findings should be interpreted by taking into consideration that cows reared in organic production systems live longer and therefore lactation numbers are higher than in conventionally reared cows (Stiglbauer et al. 2013; Orjales et al. 2016; Villar et al. 2016). It must also be taken into account that parity number strongly influences SCCs (Renau 1986). In fact, the SCC logarithm increases gradually with parity, from 4.85 in first lactation to 5.35 ($\geq$ 7 lactations). The increase in SCC with lactation number has been studied in both organic and conventional systems, and it is higher in organic systems (Orjales et al. 2017). Milk production is another important factor that influences SCCs and is attributed to a dilution effect (Green et al. 2006) (Fig. 7.1).

Farms that avoid using antibiotic treatments must improve sanitary measures to reduce SCCs, as previously reported data indicate that strategies for reducing SCCs in organic farms are ineffective (Orjales et al. 2016). The increase in SCCs in organic cattle that are not given antibiotic treatments as lactation number increases leads to a correlation between the increased SCCs and chronic infection as a result of avoiding the use of antibiotics (Orjales et al. 2016; Villar et al. 2016). Older cows

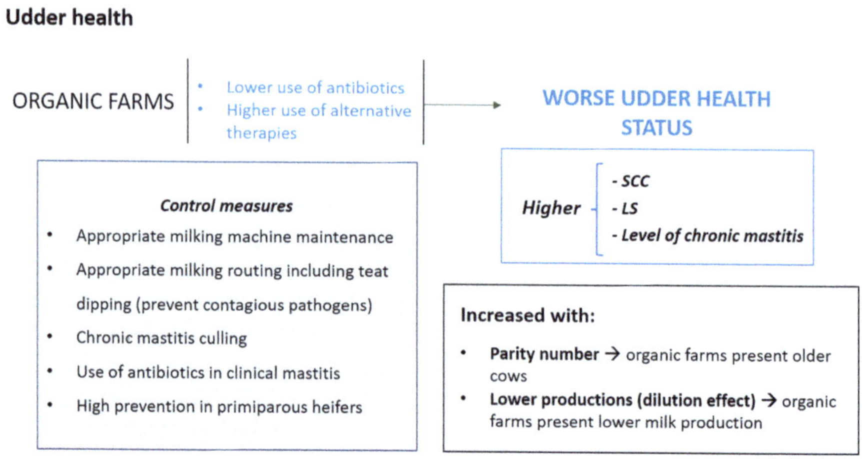

Fig. 7.1 Factors affecting SCC, LS and chronic mastitis in organic systems. Control measures to prevent the increase of mastitis disease

are more likely to suffer from chronic mastitis (higher SCCs) because of the accumulated risk of having intramammary infections that are not totally cured, resulting in a gradual increase in the number of infected quarters (Renau 1986) and a decrease in the rate of successfully cured infections after several antibiotic treatments (Swinkels et al. 2013).

The chronicity of udder infections may be related to restrictions in the prophylaxis and treatment (particularly dry cow therapy) of cases of mastitis (Vaarst et al. 2006). However, other authors (Bennedsgaard et al. 2010) suggest that antibiotic treatments for udder infection may be reduced without any apparent negative effects and the control measures for SCCs used in organic farms are at least as effective as those used for this purpose on conventional farms. Nonetheless, although preventive practices implemented on all dairy farms are of vital importance, they are particularly important on organically managed farms because of the limited availability of products to treat mastitis (Stiglbauer et al. 2013).

Organic regulations limit antibiotic use, and many organic producers prefer to administer alternative treatments to manage mastitis, but which increase SCC. In other words, udder health status is poorer on farms that use alternative therapies than on farms that use antibiotic therapies. This situation is aggravated by parity number in farms that avoid using antibiotic treatments, and the use of antibiotic therapies in older cows suffering from mastitis is highly recommend to maintain udder health (Orjales et al. 2017). Older cows are at greater risk of suffering chronic subclinical mastitis, which is very difficult to control without antibiotic therapy (Cicconi-Hogan et al. 2013).

As in conventional farming systems, udder management is of vital importance in organic dairy production. Routine teat dipping contributes to preventing increments in SCCs (Orjales et al. 2017). Teat dipping is considered an effective method of reducing subclinical mastitis (Erskine and Eberhart 1991), although it is not equally effective against all bacteria (Smith and Hogan 1993). It is more effective in preventing infection by contagious pathogens than infection by environmental streptococci (Hogan et al. 1989). As contagious pathogens are usually more important than environmental pathogens in organic dairy farming (because of the prohibition of use of blanket dry cow therapy) (Villar et al. 2016), teat dipping may be an effective means of control.

Prevention of mastitis begins with appropriate care of heifers. A recent study published by Peña-Mosca et al. (2023a) identified *Staphylococcus aureus*, *Staphylococcus chromogenes* and *Streptococcus* spp. as the most prevalent microorganisms causing mastitis in organically reared primiparous cows during the first month of lactation. These microorganisms were identified in respectively 46.2%, 48.1% and 50% of primiparous dairy cows on organic farms, showing a high persistence of intramammary infection. The persistence of intramammary infection explains why the high SCCs persisted throughout and across lactations (Fernandes et al. 2021). *Streptococcus* spp. were associated with lower milk production in the following lactation (Valckenier et al. 2020), which indicates the potential importance of controlling these bacteria in organically reared animals (Peña-Mosca et al. 2023b). Farmers must take particular care to control infections in heifers because of the suggested predominance of udder-adapted mastitis pathogens in first-lactation organic dairy cows, which may impact udder health and performance during subsequent lactations (Peña-Mosca et al. 2023b).

As antibiotic treatments are strongly restricted in organic dairy systems, prevention is essential to control udder infection (Fig. 7.1). Both organic farms and conventional farms converting to organic should take the following recommendations into account: older cows with high SCCs must be eliminated, antibiotic treatments should be used at least in cows with high SCCs (clinical mastitis) and an appropriate milking routine including teat dipping should be used (Orjales et al. 2017).

7.3.2 Helminthiases

Controlling parasitic infections is an important concern in organic farming, as grazing may expose animals to high parasite loads (Wagner et al. 2021). Helminthiases are among the most common parasitic infections in pasture-based systems and have a detrimental effect on feed intake, growth, productivity, reproductive standards and the health and welfare of dairy cattle (Rose Vineer et al. 2020). Adult dairy cattle usually have subclinical helminthiases that cause significant production losses (Mezo et al. 2011). The most pathogenic species of helminths in Europe are the stomach worm *Ostertagia ostertagi* (Forbes et al. 2008), the lungworm *Dictyocaulus viviparus* (Bloemhoff et al. 2015) and the liver fluke *Fasciola hepatica* (Mezo et al. 2008).

Control of helminthic parasites in livestock has traditionally been via administration of parasiticides. However, this approach is increasingly unsustainable as anthelmintic resistance is spreading and is predicted to increase further in the near future

(Rose Vineer et al. 2020). Control of helminths in dairy cattle is especially difficult because deworming is only permitted during the dry period, which occurs consecutively on farms throughout the year. Treated and untreated cows may share the same pastures, and as untreated ruminants continue to shed eggs in the faeces, there is a higher risk that all of the cows will ingest infective stages of helminths while feeding on grass (Voinot et al. 2020). However, some authors consider that tolerance to parasites may be higher in organic systems than in conventional farms, because of restrictions on use of anthelmintics. Thus, the host-parasite relationship may be modified by exposure of the livestock to parasite challenge from an early age, which may lead to improved resilience (Keatinge et al. 2002) (Fig. 7.2). Nonetheless, correct control of parasite infections is essential to achieve the underlying goals of organic milk production and to improve both animal welfare and product quality.

Although levels of parasitic infection may be higher on organic than on conventional farms, recent studies suggest that, overall, infections levels are similar both types of farm (Orjales et al. 2017). In fact, while more than 70% of organic herds and 40% of animals were exposed to *Fasciola hepatica*, only 54.5% of the herds and 13.3% of animals were demonstrated to have active *F. hepatica* infections (Fig. 7.2). However, the high level of exposure observed on organic farms may explain milk

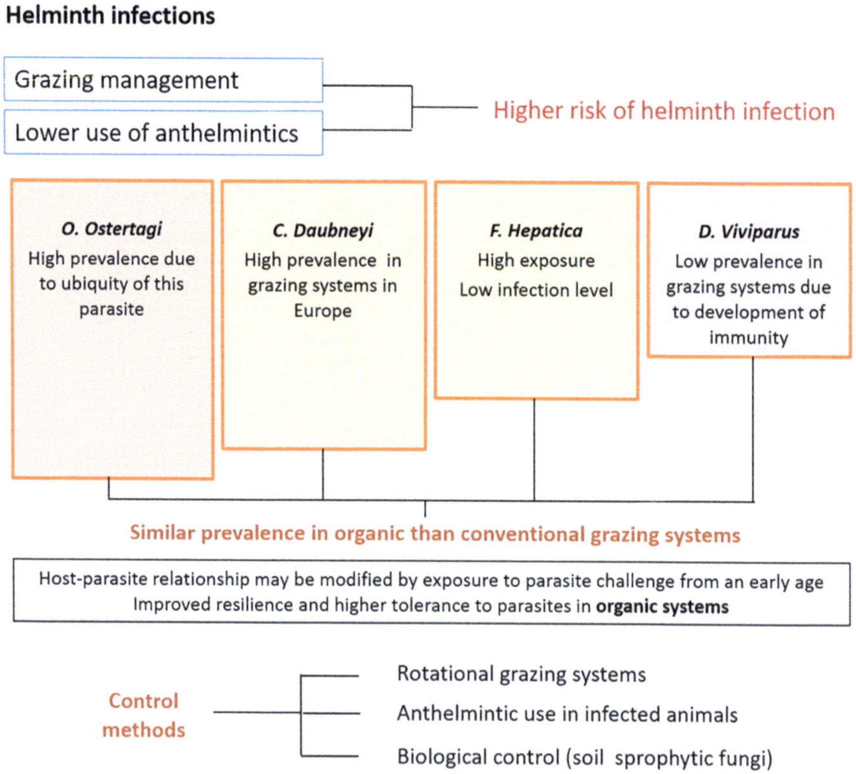

Fig. 7.2 Helminth infections: prevalence, risk of infection and control methods

production losses (Mezo et al. 2011). The high proportion of exposed herds demonstrate that fasciolosis is widespread in organic farms, mainly in regions with favourable climate conditions (high humidity, frequent rainfall and mild temperatures) that provide a suitable environment for development of the intermediate host *Galba truncatula* (Mezo et al. 2008).

In the case of the trematode *Calicophoron daubneyi*, researchers have observed that the prevalence of this parasite was higher in organic systems (herd prevalence = 59.1% and individual prevalence = 33.9%) than in conventional systems in the same region of Spain (González-Warleta et al. 2013; Ferreras et al. 2014) (Fig. 7.2). However, these results can be attributed to the increasing prevalence of *C. daubneyi* in several European countries rather than only to the effect of the production system type (Toolan et al. 2015).

The nematode worm *Ostertagia ostertagi* is prevalent on organic dairy farms, as demonstrated in studies conducted in Spain (Orjales et al. 2017) and Sweden (Höglund et al. 2010) (Fig. 7.2). However, high prevalence rates are also observed on conventional dairy farms due to the ubiquity of this parasite in temperate climates (Pablos-Tanarro et al. 2013). Nevertheless, elevated levels of this parasite have been associated with significant productivity losses (Forbes et al. 2008).

Prevalence of the parasitic nematode *Dictyocaulus viviparus* is very low in organic herds, as well as in conventional pastured-based herds (Orjales et al. 2017; Ploeger et al. 2012) (Fig. 7.2). This is because organic animals are raised on pasture, allowing adult cows to develop immunity to the parasites after experiencing the infection as calves (Ploeger et al. 2012) and thereby preventing the development of patent infections.

Anthelmintic treatments, lactation number and grazing management are among the most important factors affecting helminth infections. While the use of anthelmintics generally reduces the risk of infection, this may not always be effective because many anthelmintics can only be administered during the dry period. In cattle raised in pasture-based conditions year-round, such as organic cattle, continuous reinfections can occur, thereby reducing the expected effectiveness of anthelmintics (Orjales et al. 2017; Mezo et al. 2008; Ferreras et al. 2014; Charlier et al. 2007).

In addition to the classical control methods used in the farms studied (grazing management and anthelmintic treatment), biological control or planting specialized crops which reduce infection level can be applied (Thamsborg et al. 2010). Saprophytic fungi can be used in biological control strategies for helminths. Some saprophytic fungi (*Duddingtonia flagrans, Monacroporium thaumassium*) are able to trap the infective stages (de Araújo et al. 2004), while others (*Pochonia chlamydosporia, Mucor circinelloides, Purpureocillium lilacinum*) have antagonistic effects on helminth eggs and on trematodes, cestodes and nematodes (Cazapal-Monteiro et al. 2015) and can also act against the miracidia of *C. daubneyi* (Arroyo et al. 2017) (Fig. 7.2).

A study on grazing cattle administered mycelium of the ovicide fungus *Pochonia chlamydosporia* for 18 months revealed a 67% reduction in the number of *F. hepatica* eggs in the faeces of the cows (Dias et al. 2013). In addition, the numbers of third-stage larvae of gastrointestinal nematodes decreased by 73.8% in the faeces of cows orally administered chlamydospores of *D. flagrans* and by 58–73.5% in pasturing zebu (Mendoza-de Gives et al. 2018). In a recent study, Voinot et al. (2020), significant reductions in the egg-output values of *C. daubneyi* and gastrointestinal nematodes were recorded after daily administration of fungal chlamydospores to dairy cattle heifers.

The traditional method of reducing the presence of infective stages in the soil, by applying rotational grazing regimes in which ruminants feed on grass for a certain length of time (grazing period) and the pasture is then allowed to rest (resting period), is advised (Undersander et al. 2002) (Fig. 7.2). The resting period may also reduce the survival of infective stages of parasites in the soil (Chandrawathani et al. 2004).

In general, levels of helminth infections in organic dairy farming are similar to or even lower than in conventional systems. Distribution of the parasites is not homogenous and is strongly influenced by environmental conditions (Chincarini et al. 2022). Nonetheless, specific sustainable control programmes, planned according to the different species circulating at the farm level, should be evaluated (Chincarini et al. 2022).

7.4 Nutrition in Organic Grazing Systems

Providing cows with appropriate nutrition is one of the main challenges in organic dairy farming (Orjales et al. 2019). Adequate nutrition ensures the health and performance of livestock, particularly in dairy cattle in which requirements for optimal production, good health and reproductive efficiency are elevated (Weller and Bowling 2007).

Pasture-based systems are generally associated with seasonal spring calving, in which the main food source during the year is the pasture directly consumed by the cows during grazing. This system favours the use of small animals and with lower production than in confinement systems, but it promotes good levels of production per area unit and robust animals. Development of "specialist grazing cows" has been promoted mainly for seasonal grazing systems and with low supplementation, suggesting a robust, autonomous and "easy care" cow (Roche et al. 2018).

In conventional dairy farming, high-yielding cows are commonly fed in excess of requirements by giving them diets supplemented with high amounts of energy and protein concentrates. However, in organic farming, the aim is to optimize use of available resources rather than maximize production (Van Wagenberg et al. 2017). The limitation of available resources usually leads to lower milk yields in organic farming than in conventional production (Leiber et al. 2017; Rodríguez-Bermúdez et al. 2019a).

As mentioned in Sect. 7.2 (European Commission 2018, 2021) (Fig. 7.3), bovines should be fed with at least 70% food produced on the farm itself. Animals must have access to pasture for grazing whenever conditions allow, and at least 60% of the DM

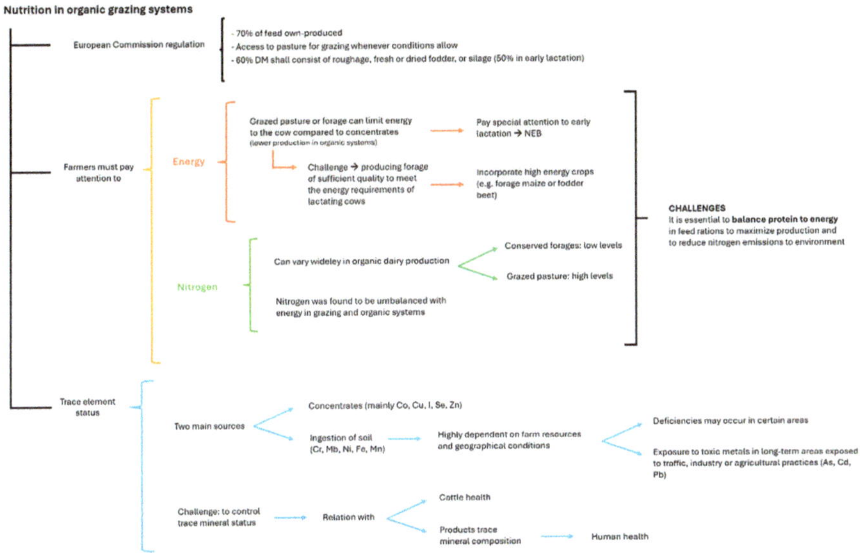

Fig. 7.3 Principle recommendations attending to nutrition in organic grazing systems. (DM Dry Matter, NEB Negative Energy Balance, Co cobalt, Cu copper, I iodine, Se selenium, Zn zinc, Cr chromium, MB molybdenum, Ni nickel, Fe iron, Mn manganese, As arsenic, Cd cadmium, Pb lead)

should consist of roughage, fresh or dried fodder, or silage. However, in the case of early lactation (maximum 3 months) this proportion can be reduced to 50%. Diets based on resources that farmers can obtain on their own farm, such as grazed-pasture or forage, are cheaper than concentrates but may be limited in the provision of energy to the cow (Blair 2021). The decrease in milk yield of organic cattle relative to conventional reared cows can be explained by the limited allowance of concentrates in organic diets (Erlt et al. 2014). In fact, milk production can be 32% lower in dairies using 75–100% pasture-based forage than in those using 25% or less pasture forage (McBride and Greene 2009), where insufficient Dry Matter Intake (DMI) of pasture by dairy cows is a limiting factor for milk production.

DMI in lactating cows largely depends on the type of feed supplied, level of feeding, ration formulation and quality of feed (Mazumder and Kumagai 2006). Orjales et al. (2019) found that the DMI was almost 25% lower in organically managed cows, in which less concentrate feed was supplied in the diet. Furthermore, the same study showed that feedstuffs used in organic farms contained more fibre, both neutral detergent fibre (NDF) and acid detergent fibre (ADF), than those used in conventional farms, so that the diets used in organic farms would be less digestible. Organic farms generally use higher proportions of forage (more than 80% of diet), and they use different types of forage with less maize silage and more hay in the diets (Orjales et al. 2019).

The most critical period for dairy cows is the periparturient period, particularly during early lactation, when the amount of energy the cows need to produce milk is greater than the amount of energy they obtain by eating, so that dairy cows in early lactation often have a negative energy balance (Hammon et al. 2009) (Fig. 7.3). Conventional farmers tend to supplement concentrates to mitigate this deficit, and the situation may therefore be more critical in grazing-based organic systems. Lipomobilization after calving is higher in organically reared cows than in conventionally reared cows, and the proportion of cows with subclinical ketosis is greater in organic than in conventional farms (Abuelo et al. 2014).

Energy intake and energy concentration of diets tend to be lower in organic farming systems (Orjales et al. 2019), which is not surprising because of the lower energy intake typical of pasture-based systems than in intensive conventional productions (Hofstetter et al. 2014). The reduction in concentrates and maize silage, which are primary sources of energy for cattle, contributes to this difference in organic diets. Given that energy is often a limiting factor in grazing-based dairy systems, incorporating high energy crops in the diets (e.g. forage maize or fodder beet) can provide energy-rich forage feed and enhance the quality of diets (Hofstetter et al. 2014). Consequently, one of the main challenges for farmers transitioning from conventional to organic systems is to produce forage of sufficient quality to meet the energy requirements of lactating cows. This challenge extends to all farmers operating in grazing conditions, particularly in organic systems in which forage comprises the main component of the diet. The quality of forage also depends on the type of forage, herbage species, the plant stage of development and the soil and climatic conditions (Blair 2021). Although forage age may be the most important factor (because the fibre is increasingly lignified over time, thus reducing the digestibility of the plant material), the methods used to preserve forage add another source of variation to the feed quality (Randby et al. 2012).

Nitrogen supply is another key factor and can vary widely in organic dairy production, ranging from low levels in rations based on conserved forage, to a surplus of protein in grazing cows (Weller and Bowling 2007) (Fig. 7.3). In the study conducted by Orjales et al. (2019), the protein concentration tended to be higher in grazing systems than in no-grazing systems. Pasture usually contains more protein than other types of forage (Hofstetter et al. 2014), although the content depends on the nitrogen supplied through fertilizers (restricted in organic farming) and also on the species cultivated.

In grazing and organic systems, the degradable nitrogen is excessive or the energy supply is limited for the synthesis of microbial protein, and therefore the dietary protein is not efficiently used (Orjales et al. 2019). The excess nitrogen is mainly excreted as urinary urea, and increased excretion of nitrogen per unit of dietary protein consumed by animals fed unbalanced diets is thus expected (Orjales et al. 2019). Unbalanced diets result in less dietary nitrogen being converted into milk protein and more nitrogen being released to the environment (Koesling et al. 2017). Balancing protein to energy in feed rations is considered one of the most effective dietary interventions that will reduce nitrogen emissions by ruminants (Eckard et al. 2010).

For farmers converting to organic systems, and even for experienced farmers in organic production, the challenges are therefore to produce sufficient, good quality forage, with the purpose of minimizing concentrate use, to meet the energy and protein requirements of the dairy cows. This is a difficult task as the conditions on each farm are different and very variable within different regions (Hardie et al. 2014). As mentioned above, the main challenge for organic farmers is to balance the diets in order to maximize production based on their own resources. The limited energy intake in organically-reared cows is associated with lower protein use efficiency. In order to match the milk production requirements more precisely and to improve protein use efficiency, thereby reducing loss of nitrogen to the environment, organic farming should aim to increase the energy content of cattle diets by improving forage quality and formulating rations with more balanced combinations of forage and grain (Orjales et al. 2019).

7.4.1 Trace Element Status

Mineral nutrition is an important challenge in organic dairy farming (Fig. 7.3). Most organic herds exhibit significant deficiencies in trace elements (Rey-Crespo et al. 2014; Orjales et al. 2018). Organic farming is highly dependent on on-farm resources and geographical conditions, and therefore mineral deficiencies may occur in some areas due to the low availability of some trace elements. For example, the distance to the sea is an important factor in determining the concentrations of iodine and to a lesser extent of selenium in the soil/pasture (Flachowsky et al. 2014).

Cattle mineral status is associated with grazing intensity, which involves ingestion of soil directly or through soil-contaminated forage (López-Alonso 2012). In fact, ingestion of soil may represent an important source of essential trace elements for livestock (Thornton and Abrahams 1983); however, it can also be an important source of exposure to toxic metals in areas exposed in the long term to traffic, industry or agricultural practices (Miranda et al. 2005). This may be another challenge in the conversion from conventional to organic systems, particularly for farmers who are used to working under no-grazing conditions in which all mineral sources are well controlled. In organic systems, mineral sources are highly dependent on the local geological conditions, and the nutritional status of grazing cattle varies throughout the year. The availability of forage may be very low during some months of the year, depending on weather conditions, leading to a strong dependence on preserved forage or concentrate feed. The trace element content of forage also varies throughout the year (Lean and Golder 2023; Linehan et al. 2024). Seasonal variations in forage trace element concentrations largely depend on geographical and climate conditions. The selenium content of plants is lower in areas with high rainfall and its uptake by plants is largely determined by the temperature (higher rate at temperatures >20 °C), whereas iron concentration in pasture undergoes seasonal fluctuations, with peaks in spring and autumn (Kabata-Pendias 2011). Trace element concentrations also depend on the stage of plant development: higher contents of iron, zinc and copper were detected in alfalfa leaves in the first stage of plant development (Marković et al. 2009), and concentrations of molybdenum and zinc are also known to vary with plant development stage (Kabata-Pendias 2011).

Preservation of forage also affects the trace element content, and for example the fermentation process appears to increase the iron bioavailability in silage (Hansen and Spears 2009). In addition, the degree to which plants are contaminated with soil can also affect the trace element concentrations, particularly in elements present at much higher concentrations in soil than in plants. Fresh forage and silage are more likely to be contaminated with soil than hay. Thus, the bioavailability of trace minerals in plants can be very variable, and nutritional requirements are therefore difficult to calculate (Orjales et al. 2018).

The lack of trace element supply in organic cattle affects the health of the cattle (reproductive disorders, immune function, energy metabolism and other diseases) (Overton and Yasui 2014; Blair 2021) and can also lead to the milk containing low levels of minerals; the altered milk quality is very important in relation to how consumers view the product (Rey-Crespo et al. 2013, 2014; Bath and Rayman 2016; Rodríguez-Bermúdez et al. 2020a; Linehan et al. 2024). Human health may also be affected and for example iodine deficiency in humans can affect foetal development and the neurological development of children. This has led to concern about the typically low iodine content of organic milk in Europe (Rey-Crespo et al. 2013; Bath and Rayman 2016; Srednicka-Tober et al. 2016).

For practical purposes trace elements requirements are oriented to the total intake of trace elements in the ration. However, evaluating trace element concentrations in different feed materials is essential to understand the sources of trace elements in organic dairy cattle (Orjales et al. 2018; Swecker 2023). In general, no differences between feed material produced in organic and conventional systems have been found, except in concentrate feed for the trace element added as mineral supplements, which were always lower in the organic systems (Orjales et al. 2018).

A recent study (Orjales et al. 2018) has demonstrated that trace mineral status of organic cattle depends on two main sources. The first source, concentrate fed to cows, includes trace elements in the mineral supplements at concentrations higher than in the other feedstuffs (mainly, cobalt, copper, iodine, selenium and zinc), thus preventing mineral deficiencies. The second source, ingestion of soil during grazing or consumption of soil contaminated forage, is the main source of trace elements that are not supplemented (chromium, molybdenum and nickel), trace elements supplemented but present at higher concentrations in soil/forage than in mineral supplements (iron, manganese), and toxic elements (arsenic, cadmium, lead).

As previously mentioned, cattle mineral status is not only related to animal health but is also important in relation to the mineral content of the products. Trace element concentrations are significantly lower in organic milk than in conventional milk, especially the concentrations of iodine, selenium, copper and zinc, which are routinely supplemented in concentrate food. As in organic dairy production, inorganic mineral supplements are not recommended, and other supplements are required to increase the mineral content of cattle diets and consequently, the mineral content of milk. Rey-Crespo et al. (2014) evaluated the use of algae as a source of minerals (especially iodine but also other trace elements) in organic dairy cattle. The algal supplements were found to significantly improve the animal mineral status and the mineral content of milk, particularly the iodine and selenium contents.

7.5 Breed Selection

During the conversion from conventional to organic systems, farmers usually maintain the same livestock. However, after the transition period they may begin to consider whether another breed of cow might be better adapted to organic conditions. In this section we will consider the advantages and disadvantages of different pure dairy breeds and crosses (Figs. 7.4, 7.5, 7.6 and 7.7).

Since the 1970s, Holstein cattle have become widely used in much of the world (Rodríguez-Bermúdez et al. 2020a). Selection of the Holstein breed has exclusively focused on improving milk yield during most of the selection process (Ahlman 2010; Davis et al. 2020), although in recent years Holstein breeders have paid more attention to other traits such as e.g. longevity, reproduction and other functional traits. The result of this intensive selection for milk yield is a highly specialized animal that is very well adapted to zero-grazing systems (Ahlman 2010; Rodríguez-Bermúdez et al. 2020a).

Despite the recommendations included in EU regulations regarding the adaptation of breeds to organic dairy production, little effort has been made to produce animals that are well adapted to organic conditions. As already mentioned, most dairy farmers maintain the same livestock when converting to organic systems (Rodríguez-Bermúdez et al. 2019a, b, 2020a), with Holstein being the predominant breed (Ivemeyer et al. 2018). However, many researchers have pointed out that Holsteins may not be the best choice for organic systems, in which the use of chemical treatments is reduced or prohibited, and grazing is particularly important (Nauta

Fig. 7.4 Pure Holstein cows in an organic farm

Fig. 7.5 Pure Brown Swiss and crosses (Holstein x Brown Swiss) in an organic farm

Fig. 7.6 Pure Holstein, pure Swedish Red and crosses (Holstein x Swedish Red and Holstein x Fleckvieh) in an organic farm

et al. 2006; Ahlman et al. 2011; Davis et al. 2020). In this respect, different authors have highlighted that Holstein cattle have reproductive problems such as reduced reproductive efficiency and extended calving intervals, or high culling rates, which contribute to shortening their productive life (Nauta 2001; Bluhm 2009).

In addition, organic producers exhibit some dissatisfaction regarding the performance of Holstein cows, demanding a more robust cow (Rodríguez-Bermúdez et al. 2019a). Despite the EU's recommendations, it is almost impossible to find animals that have been bred specifically for organic production systems, and farmers usually turn to 'tried and trusted' conventional breeds (Slagboom et al. 2016).

From a scientific point of view, the genotype by environment interaction (GxE) plays an important role in production development (Van Diepen et al. 2007). One

Fig. 7.7 Organic herd originated from multiple crossed cows obtained from Holstein, Jersey, Montbeliarde, Swedish Red, Fleckvieh and Brown Swiss

example of the effect of such interactions is the reduction in the reproductive and productive performance of highly productive American Holstein cows in low-input systems, which occurs because these cows cannot express their full genetic potential in the absence of supplementation. Studies comparing different Holstein Friesian strains, New Zealand (NZ) and North American (NA) breeds, with similar genetic merit for milk traits, have shown that under grazing conditions NZ cows consume a greater amount of pasture and produce more milk solids, expressed as a percentage of their body weight, and also had a lower mobilization of body reserves and better fertility than NA cows. However, when provided a total mixed ration (TMR) diet, the NA cows performed better (with higher DMI and milk production) (Baudracco et al. 2010).

The existence of strain × environment interactions suggests that the most appropriate strain of cow will differ depending on the production system under which it is managed. Breeding plans should therefore consider individualized system-level breeding objectives (Baudracco et al. 2010).

For these reasons, having cattle better adapted to the specific production system would result in lower production costs and enhanced production efficiency (Simm et al. 2004). Recent studies estimate that the main challenge to sustainable organic dairy farming is to identify the cattle genotypes that are best adapted to forage-based production systems (Peeters and Wezel 2017).

In this context, some authors have studied the performance of different pure breeds (including Holstein) and their crosses, by analyzing milk yield production, milk solids production, udder health, reproduction and many functional traits under organic conditions. The interest shown in crossbred cattle is due to the potentially increased productivity due to heterosis or hybrid vigour, i.e. when crosses perform better for one or more characters than both parental breed (Rodríguez-Bermúdez et al. 2019b).

7.5.1 Selection for Milk Yield and Udder Health

Regarding milk yield, most researchers found that Holstein cattle produce higher milk yields than both dairy crosses and other pure-breed cattle (de Haas et al. 2013; Rodríguez-Bermúdez et al. 2017, 2020a; Bieber et al. 2020). In general, crosses produce milk yields that are intermediate between those of both Holstein and the other pure breeds included in the cross (de Haas et al. 2013).

Some researchers have demonstrated that under grazing conditions Holstein cows supplemented with moderate amounts of concentrate yielded maximum economic profits (Baudracco et al. 2010). This result may explain why some organic farmers prefer to continue using Holstein cows (and provide larger amounts of supplements than other farmers) and not use local breeds (Nauta 2001). The best example of Holstein cows adapted to pasture-based systems are rustic Friesians in New Zealand, where selection prioritizes milk solids yield and fertility at the expense of milk yield (LIC 2024). However, the New Zealand Holstein breed may not be the best option for farmers who sell their milk in payment systems that mainly consider volume, as the cows are selected to produce large quantities of milk solids. Organic farmers who prefer to use rustic Holsteins should select their own Holstein strain focusing on improving reproduction, functional traits and disease resistance while maintaining good milk yield production (Rodríguez-Bermúdez et al. 2019b).

Milk solids (fat and protein) production is determined by both breed and milk yield. De Haas et al. (2013) found that other dairy breeds yield milk with higher protein and fat contents than Holstein cows. Similar results were obtained in organic farms in North Spain where Holstein cows tend to produce high milk yields but with significantly lower fat and protein contents than other breeds or crosses (Rodríguez-Bermúdez et al. 2017, 2020a). Similar results were also obtained in studies conducted in Germany and Sweden (Bieber et al. 2020). In fact, fat and protein yield usually is higher in crosses than in the parental breeds (Holstein x other dairy breeds), probably because of a combination of intermediate milk yield production and higher percentages of fat and protein (Rodríguez-Bermúdez et al. 2020a), which can be attributed to hybrid vigour.

Regarding the influence of breed on udder health, variable results have been obtained. In general, SCCs are more strongly influenced by the management system than by the breed type (Orjales et al. 2016; Rodríguez-Bermúdez et al. 2017). Some studies did not observe any difference in SCCs between Holstein and other dairy breeds (Horn et al. 2012; Mullen et al. 2015; Bieber et al. 2020). In other cases, SCCs were statistically significantly higher (Sundberg et al. 2009; Coffey et al. 2016) or lower (Swalve 2007) in Holstein than in other breeds or crosses.

For example, Berry et al. (2007) and de Haas et al. (2013) found that SCCs were higher in Jersey cow milk than in milk from other breeds. This could be explained by a dilution effect (Villar and López-Alonso 2015), as Jersey cows tend to produce less milk than other breeds. Swalve (2007) and de Haas et al. (2013) reported that SCCs were lower in crosses than in purebred Holstein.

7.5.2 Selection for Reproductive Efficiency

Reduced ability to reproduce leads to higher culling because of infertility, which is particularly important in organic systems where the use of oestrus synchronization protocols is prohibited and reproductive treatment options are highly restricted (EFSA 2009). As organic farming is a pasture-based system, seasonal planning of calving would be useful (Auldist et al. 2007), and good reproductive performance is of vital importance.

Regarding reproduction, crosses and other pure breeds are more fertile than Holstein cows (Swalve 2007; Sundberg et al. 2009; de Haas et al. 2013; Rodríguez-Bermúdez et al. 2019b, 2020a; Bieber et al. 2020), and crosses perform even better than both parental breeds (Holstein and other dairy breeds) (Rodríguez-Bermúdez et al. 2020a). The different reproductive performance of crosses and other dairy pure breeds relative to Holstein cows reared on the same organic farm may be related to differences in expression of oestrus, as Holstein cattle has a less marked oestrus expression than other breeds (Rodríguez-Bermúdez et al. 2019a). Dairy breeds other than Holstein also have more obvious signs of oestrus and strong sexual and mounting activity (Sveberg et al. 2015). The longer duration of the oestrus period and the better performance regarding oestrus signs improve heat detection and reproductive performance of crosses and other pure breeds.

The better reproductive performance of cross-bred cows may be an example of heterosis or hybrid vigour. The use of cross-bred cattle may be an effective way of counteracting the negative consequences that selection programs based on milk production have on reproductive efficiency (Buckley et al. 2014).

Reproduction is highly correlated with milk yield production, and the poorer performance of Holstein cows regarding reproduction may be explained by selection for higher milk yield production (Bluhm 2009; Rodríguez-Bermúdez et al. 2020a). For example, Horn et al. (2012) studied a Holstein strain selected for improved life productivity and fertility at the expense of milk yield, observing that the less productive Holstein cattle were even more efficient than Brown Swiss in terms of reproduction. It is therefore possible to produce milk under organic conditions by using Holstein cows, but some effort should be made to improve reproductive efficiency by awarding a higher score on indexes to reproductive traits at the expense of milk yield (Rodríguez-Bermúdez et al. 2019a, b).

Apart from reproductive efficiency, reproduction also considers other characters of interest, such as calving type and calving ease. In dairy production, twinning is not desirable because of some negative impacts (EFSA 2009) as a higher risk of being culled of those cows giving birth to twins (Thomsen et al. 2007) and the lower rate of replacement of heifers owing to death of twin calves or the presence of freemartins (Nielsen et al. 1989).

The natural rate of twinning is around 1–5% (Sreenan and Diskin 1989). In a recent study of organic dairy conditions, all breeds tested (Holstein, Brown Swiss and dairy crosses) showed a standard rate of twinning (Rodríguez-Bermúdez et al. 2020a). These findings differ from observations in conventional systems in which

Holstein cows tend to have higher rates of twinning than other breeds (Çobanoglu 2010). The lack of differences between breeds in organic systems may be due to fewer differences in milk production of Holstein relative to other dairy breeds or crosses, as differences in milk yields are greater in conventional systems (Rodríguez-Bermúdez et al. 2017). Another possible explanation for the lower rates of twinning in organic systems is the prohibition of the use of reproductive treatments for synchronization, as it is known that reproductive treatments contribute to higher twinning rates (EFSA 2009).

Calving ease is also of economic importance on dairy farms as it is related to calf mortality (Lombard et al. 2007), lower milk production and poor health (Heins et al. 2006). Some studies have reported benefits of cross-breeding in reducing difficulties associated with calving (Heins et al. 2006), reporting unfavourable heterosis for direct effects on calving ease and favorable heterosis for maternal effects (Sorensen et al. 2008). Holstein and Brown Swiss cows apparently need more assistance during calving than dairy crosses (Rodríguez-Bermúdez et al. 2020a). Few studies have considered this variable, and calving ease can be assumed to be similar in all pasture-based systems, independently of whether the farm is organic or not. Dhakal et al. (2013) and Mullen et al. (2015) reported no differences between Holstein, Jersey and crosses regarding calving ease.

However, management systems regarding delivery have a greater impact on calving ease than breed; on small and medium size farms cows are systematically given assistance during calving, while 97% of cattle on large farms do not need assistance (Rodríguez-Bermúdez et al. 2020a).

7.5.3 Selection for Functional Traits

Reasons for culling are usually classified as voluntary (low productivity and farm management reasons) and involuntary (illness, infertility or death). Although involuntary culling is necessary to reduce suffering in individual animals, high levels of this type of culling can have an economic impact on farms (Ahlman et al. 2011). There is little information available about culling reasons in organic systems.

Rozzi et al. (2007) found that in Canada, where Holstein is the predominant cattle breed, fertility and mastitis are the most important reasons for culling. Similar results have been observed in Sweden in a comparison of Holstein and Swedish Red breeds (Ahlman et al. 2011), although both breeds showed similar pattern, with a lower proportion of animals in Holstein group culled because of poor fertility.

In a recent study carried out on organic farms in Northern Spain (Rodríguez-Bermúdez et al. 2020a), the various different reasons for culling in dairy breeds and crosses were investigated. Holstein cows were mainly culled for multiple reasons (various accumulated pathologies), which were also important in the cross-bred and Brown Swiss cows. These findings are consistent with previous reports by organic farmers that cows were mainly culled because of advanced age (Rodríguez-Bermúdez et al. 2016) and suggest that in a relatively high proportion of animals there was no single main reason for culling cows (Rodríguez-Bermúdez et al. 2020a). In the same study, Holstein cows were found to suffer less frequently from mastitis and infertility than the cross-bred or Brown Swiss cows. The probability of Holstein cows suffering multiple pathologies but maintaining an acceptable milk yield is higher than for other breeds or crosses, in which a single main reason for culling was usually identified (Rodríguez-Bermúdez et al. 2020a).

A higher percentage of Holstein tend to be culled because of urgent slaughter than crosses or Brown Swiss cows (Rodríguez-Bermúdez et al. 2020a), which has been associated with a negative energetic balance in the post-partum period (Fouz et al. 2014).

The lower incidence of lameness in the crosses than in Brown Swiss and Holstein cows (Rodríguez-Bermúdez et al. 2020a) may be related to a lower incidence of clinical lameness in crosses than in both parental breeds (Chawala 2011) due to heterosis. This is an important observation because cows reared in pasture–based organic systems must have robust legs and hooves (Rodríguez-Bermúdez et al. 2020a).

Longevity is an important functional trait described as the capacity of a cow to avoid being culled due to low production, low longevity or illness. Longevity is therefore closely related to culling management (Vollema and Groen 1996). While cows can potentially live for more than 20 years, they generally do not reach 6 years of age on most modern dairy farms (Rushen and Passillé 2013). The decline in longevity of cattle in dairy farming generally results from involuntary culling, which clearly reduces the profitability of dairy farms and does not match the sustainability that is one of the principles of organic farming (Ahlman et al. 2011; Rushen and Passillé 2013).

As reported in a recent review (Rodríguez-Bermúdez et al. 2019b), some authors have found that crosses and other pure breeds tend to live longer than Holstein cows. By contrast, Rodríguez-Bermúdez et al. (2020a) found that Holstein cows live longer than other breeds, which is consistent with observations on organic farms in Sweden, where the longevity of Holstein and Swedish Red cattle was similar (Ahlman 2010).

7.5.4 GxE and Organic Breeding Indexes

The GxE interaction can be defined as a change in the response of genotypes to different environments or to changes in the relative merit of genotypes in the different environments. GxE interactions are especially important when animals are reared under specific environmental conditions (such as organic production). When animals are genetically adapted to particular conditions, they will be more productive and production costs will be lower (Charlesworth and Hughes 2000).

There is a GxE interaction between organic and conventional dairy production. Both GxE interactions and differences in breeding goals reduce the genetic correlation between breeding goals in both types of systems, which is an important indicator of the need for separate breeding programs for different environments (Slagboom et al. 2020).

As explained above, environment plays an important role in productive results, and the same cow could be a very good producer under intensive conventional conditions but less productive under organic conditions, owing to the interaction between the genotype and the environment (Nauta et al. 2006). The GxE interaction has been found to lead to a reranking of Holstein cows regarding productive and reproductive traits, in organic and conventional systems, indicating that some Holstein cows are better suited to organic systems (Nauta et al. 2006; Ahlman 2010). Holstein that are better adapted to organic systems are specific lines of Friesian populations, e.g. New Zealand Friesians or other rustic Friesian (Rodríguez-Bermúdez et al. 2019b, 2020a).

Access to organic breeding indexes may be useful for farmers. Many attempts have been made since researchers have pointed out that organic systems have different breeding goals than in conventional systems (Endendijk and Baars 2001; Haas and Bapst 2004; Brotherstone and Goddard 2005; Rozzi et al. 2007), some more successful than others. These indexes were thought to enable farmers to identify the conventional sires best adapted to organic production (Rodríguez-Bermúdez et al. 2019b). A further attempt has recently been made in Spain, and the CONAFE (Confederación de Asociaciones de Frisona Española) has created an index for organic dairy farms using Holstein cows (IM€T ECO) (Charfeddine and Pérez-Cabal 2019; CONAFE 2024). The scarcity of breeding indexes adapted to organic conditions is due to the fact that organic producers form a minor part of the large dairy sector. Maintaining bulls exclusively for use on organic farms is therefore not profitable. One possible solution to this problem is to adopt a scheme similar to that used by CONAFE, where bulls provided to organic farmers are the same as those provided conventional producers but they are reranked for farmers with different objectives (CONAFE 2024).

The complexity of GxE and elaborating organic breeding indexes has led some researchers to propose creative solutions. Thus, taking into consideration the high variability in environments, objectives and general conditions on organic farms, Davis et al. (2020) proposed that farmers should have access to an interactive flow chart to guide them through breed selection based on inputs constraints, and priorities within their systems, resulting in an indexing system unique to each farm.

7.5.5 *Selecting an Adapted Breed for Farm Productive Objectives*

Regarding breed selection, the first step is for farmers to analyze the existing conditions and to choose the breed or cross that adapts best to the farm environment and farming objectives (Savory et al. 2019), while also considering the capacity of the animal to adapt to local conditions and resist diseases (TEAGASC 2023) (Fig. 7.8).

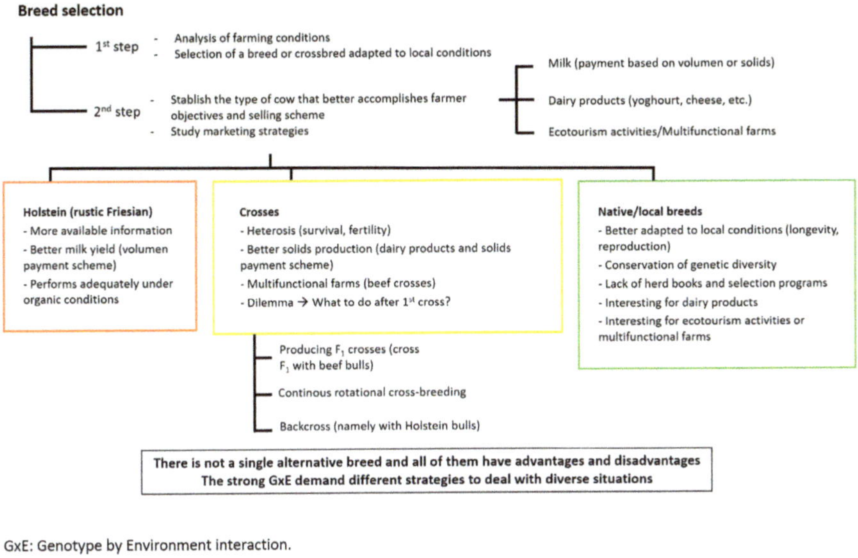

Fig. 7.8 Breed selection in organic dairy systems. Three different alternatives: rustic Friesian, Crosses and Native/Local Breeds. (GxE: Genotype by Environment interaction)

Once the type of cow that is best adapted to the particular environment has been determined, it is important to establish the type of animal that performs best according to the farmer's objectives (Rodríguez-Bermúdez et al. 2019b). The farmer should first analyze the type of product to be sold (milk volume, dairy products, tourism activities) (Rodríguez-Bermúdez et al. 2019b, 2020b; TEAGASC 2023). Available information for organic systems indicates that rustic breeds are best adapted to local conditions, displaying greater longevity and better reproductive performance than Holstein cows. The contribution to conservation of diverse genetic resources that forms part of organic philosophy (Bieber et al. 2020) could be a major strength, although the possible lack of herd books and selection programs for local breeds and crosses may hamper the selection procedure. Although crosses may not be the most productive in terms of volume, they may be valuable for yielding some dairy products because of higher milk solids production (Rodríguez-Bermúdez et al. 2019b). Multifunctional farms, producing meat (from culled cows or male calves) or participating in eco-tourism activities (Nauta et al. 2009; Rodríguez-Bermúdez et al. 2020b), could benefit from having dual purpose cows (Nauta et al. 2009; Rengab-Genetics 2013). Farmers interested in ecotourism activities should consider marketing strategies and should be aware that organic consumers prefer autochthonous breeds (Rodríguez-Bermúdez et al. 2020b).

In the case of crosses, a recent study (Clasen et al. 2020) pointed out that the main benefits of heterosis are expressed in fertility traits and survival, which ensured that fewer replacement heifers could be maintained and that the replacement rate was lower. The improved fertility in the herds permits additional use of beef semen, producing slaughter calves with a higher value. Crosses can also be beneficial for multifunctional farms or farmers undertaking milk transformation. Farmers who sell milk by volume must consider the milk payment scheme when selecting the particular breed to use. Crossbred cows may be beneficial under a milk payment scheme that rewards higher components in contrast to systems in which the payment is based on milk volume and in which breeds that maintain milk volume are more profitable, e.g. a rustic Holstein (Bluhm 2009; Rodríguez-Bermúdez et al. 2019b).

The main disadvantage of crossing is what to do after the first cross. One possible solution is to produce only F_1 crosses, which would involve maintaining part of the herd as pure-bred cows (normally Holstein). Other options include the use of continuous rotational cross-breeding strategies or even producing composite breeds (Rodríguez-Bermúdez et al. 2019b).

Backcrossing animals, in this case with a pure-bred Holstein bull, is another option (Mancuso 2017). However, backcrossing animals does not always produce good results, e.g. backcrossing Holstein x Brown Swiss with a Brown Swiss bull resulted in lower milk yields (Dechow et al. 2007). However, Mancuso (2017) found that backcrossing different F_1 crosses with Holstein bulls can recover traits of interest for milk yield.

In summary, careful evaluation of farming objectives, environmental conditions and marketing strategies before and during selection of the most suitable breed are highly recommended (Savory et al. 2019). There is no single alternative breed and all have advantages and disadvantages; different strategies are required to deal with very diverse situations because of the strong GxE interaction (Rodríguez-Bermúdez et al. 2019b).

7.6 Conclusions

Conversion from conventional to organic conditions, even when involving farms with grazing systems that are very similar to those used on organic farming, leads to a series of challenges. Three main aspects that particularly affect the organic sector are considered in this review chapter: treatment reduction, nutrition in organic breeding systems and breed selection.

As conventional treatments are restricted in organic dairy systems, farmers should pay particular attention to disease prevention. Regarding udder health, older cows with high SCCs must be eliminated, antibiotic treatments should be used in cows with high SCCs and appropriate milking routines should be established, in order to prevent mastitis. In the case of parasitosis, appropriate helminth control is essential, for which biological control (saprophytic fungi) can be used in addition to the classical control methods (grazing management and anthelmintic treatment). Nutrition is a great challenge in the organic farming sector, especially during the period of conversion from conventional to organic management. Producing sufficient good quality forage requires high training in pasture management, which is a difficult task as the conditions on different farms depend on the region considered. The big issue is to be capable of balancing diets with a view to maximize production consuming farm resources.

The high proportion of forage in the diets and the absence of trace element supply in organic dairy farm systems can affect cattle health and lead to low mineral contents in the milk produced. Natural products such as algae can be included in the diet as a source of microminerals in dairy cattle.

Finally, before selecting a breed, farmers should carefully analyze the environmental conditions on their farms, as well as their farming objectives, opportunities to sell their products and marketing strategies. They should have a holistic view of their production context and consider the advantages and disadvantages of all breed alternatives, and they should be aware that there may be more than a single breed solution.

Acknowledgements Ruth Rodríguez-Bermúdez, Marta Miranda, Marta López-Alonso and Inmaculada Orjales belong to the "Grupo de Potencial Crecimiento", funded by GAIN (Axencia Galega de Innovación) (grant number ED431B 2023/008).

References

Abuelo A, Hernández J, Benedito JL, Castillo C (2014) A comparative study of the metabolic profile, insulin sensitivity and inflammatory response between organically and conventionally managed dairy cattle during the periparturient period. Animal 8:1516–1525. https://doi.org/10.1017/S1751731114001311

Aerni P (2023) COP-27: a great opportunity to address the double crisis of food security and climate change–and for the EU to re-align its farm to fork strategy. Front Environ Econ 1. https://doi.org/10.3389/frevc.2022.1082869

Ahlman T (2010) Organic dairy production. Herd characteristics and genotype by environment interactions. Uppsala Sveriges lantbruksuniv, Uppsala

Ahlman T, Berglund B, Rydhmer L, Strandberg E (2011) Culling reasons in organic and conventional dairy herds and genotype by environment interaction for longevity. J Dairy Sci 94:1568–1575. https://doi.org/10.3168/jds.2010-3483

Ali AKA, Shook GE (1980) An optimum transformation for somatic cell concentration in Milk. J Dairy Sci 63:487–490. https://doi.org/10.3168/jds.S0022-0302(80)82959-6

Arroyo FL, Hernández JÁ, Cazapal-Monteiro CF, Pedreira J, Sanchís J, Romasanta Á, Sánchez-Andrade R, Paz-Silva A, Arias MS (2017) Effect of the filamentous fungus Mucor circinelloides on the development of eggs of the rumen fluke Calicophoron daubneyi (Paramphistomidae). J Parasitol 103:199–206. https://doi.org/10.1645/16-76

Auldist MJ, Pyman MFS, Grainger C, Macmillan KL (2007) Comparative reproductive performance and early lactation productivity of Jersey x Holstein cows in predominantly Holstein herds in a pasture-based dairying system. J Dairy Sci 90:4856–4862. https://doi.org/10.3168/jds.2006-869

Bath SC, Rayman MP (2016) Trace element concentration in organic and conventional milk: what are the nutritional implications of the recently reported differences? Br J Nutr 116:3–6. https://doi.org/10.1017/S0007114516001616

Baudracco J, Lopez-Villalobos N, Holmes CW, Macdonald KA (2010) Effects of stocking rate, supplementation, genotype and their interactions on grazing dairy systems: a review. N Z J Agric Res 53:109–133. https://doi.org/10.1080/00288231003777665

Bennedsgaard TW, Klaas I, Vaarst M (2010) Reducing use of antimicrobials—experiences from an intervention study in organic dairy herds in Denmark. Livest Sci 131:183–192. https://doi.org/10.1016/j.livsci.2010.03.018
Berry DP, Lee JM, Macdonald KA, Stafford K, Matthews L, Roche JR (2007) Associations among body condition score, body weight, somatic cell count, and clinical mastitis in seasonally calving dairy cattle. J Dairy Sci 90:637–648. https://doi.org/10.3168/jds.S0022-0302(07)71546-1
Bieber A, Wallenbeck A, Spengler Neff A, Leiber F, Simantke C, Knierim U, Ivemeyer S (2020) Comparison of performance and fitness traits in German Angler, Swedish Red and Swedish Polled with Holstein dairy cattle breeds under organic production. Animal 14:609–616. https://doi.org/10.1017/S1751731119001964
Blair R (2021) Nutrition and feeding of organic cattle, 2nd edn. Cab International Reading
Bloemhoff Y, Forbes A, Good B, Morgan E, Mulcahy G, Strube C, Sayers R (2015) Prevalence and seasonality of bulk milk antibodies against Dictyocaulus viviparus and Ostertagia ostertagi in Irish pasture-based dairy herds. Vet Parasitol 209:108–116. https://doi.org/10.1016/j.vetpar.2015.01.021
Bluhm W (2009) The role of crossbreeding in UK dairy breeding, pp 1–38
Brook L, Johnson A (2022) Grow organic: the climate, health, and economic case for expanding organic agriculture. https://www.nrdc.org/resources/grow-organic-climate-health-and-economic-case-expanding-organic-agriculture. Accessed 28 Jan 2024
Brotherstone S, Goddard M (2005) Artificial selection and maintenance of genetic variance in the global dairy cow population. Philos Trans R Soc Lond Ser B Biol Sci 360:1479–1488. https://doi.org/10.1098/rstb.2005.1668
Buckley F, Lopez-Villalobos N, Heins BJ (2014) Crossbreeding: implications for dairy cow fertility and survival. Animal 8:122–133. https://doi.org/10.1017/S1751731114000901
Caraviello DZ, Weigel KA, Shook GE, Ruegg PL (2005) Assessment of the impact of somatic cell count on functional longevity in Holstein and Jersey cattle using survival analysis methodology. J Dairy Sci 88:804–811. https://doi.org/10.3168/jds.S0022-0302(05)72745-4
Cazapal-Monteiro CF, Hernández JA, Arroyo FL, Miguélez S, Romasanta Á, Paz-Silva A, Sánchez-Andrade R, Arias MS (2015) Analysis of the effect of soil saprophytic fungi on the eggs of Baylisascaris procyonis. Parasitol Res 114:2443–2450. https://doi.org/10.1007/s00436-015-4440-0
Chandrawathani P, Jamnah O, Adnan M, Waller PJ, Larsen M, Gillespie AT (2004) Field studies on the biological control of nematode parasites of sheep in the tropics, using the microfungus Duddingtonia flagrans. Vet Parasitol 120:177–187. https://doi.org/10.1016/j.vetpar.2003.12.014
Charfeddine N, Pérez-Cabal MA (2019) Anexo X. Estudio Económico Para La Actualización Del Ico Y La Definición De Nuevos Índices De Mérito Económico Total Programa Cria La Raza Frisona, pp 144–147
Charlesworth B, Hughes K (2000) The maintenance of genetic variation in life-history traits. In: Singh R, Krimbas C (eds) Evolutionary genetics. Cambridge University Press, Cambrigdge, pp 369–392
Charlier J, Duchateau L, Claerebout E, Vercruysse J (2007) Predicting milk-production responses after an autumn treatment of pastured dairy herds with eprinomectin. Vet Parasitol 143:322–328. https://doi.org/10.1016/j.vetpar.2006.08.015
Chawala AR (2011) The effect of breed and crossbreeding on the incidence of recorded lameness in the New Zealand dairy cattle. Massey Universtiy (New Zealand)
Chincarini M, Lanzoni L, Di PJ, Morelli S, Vignola G, Paoletti B, Di CA (2022) Animal welfare and parasite infections in organic and conventional dairy farms: a comparative pilot study in Central Italy. Animals 12:1–13. https://doi.org/10.3390/ani12030351
Cicconi-Hogan K, Gamroth M, Richert R, Ruegg P, Stiglbauer K, Schukken Y (2013) Associations of risk factors with somatic cell count in bulk tank milk on organic and conventional dairy farms in the United States. J Dairy Sci 96:3689–3702

Clasen JB, Fikse WF, Kargo M, Rydhmer L, Strandberg E, Østergaard S (2020) Economic consequences of dairy crossbreeding in conventional and organic herds in Sweden. J Dairy Sci 103:514–528. https://doi.org/10.3168/jds.2019-16958

Çobanoglu Ö (2010) Twinning in cattle: desirable or undesirable? J Biol Environ Sci 4:1–8

Coffey EL, Horan B, Evans RD, Berry DP (2016) Milk production and fertility performance of Holstein, Friesian, and Jersey purebred cows and their respective crosses in seasonal-calving commercial farms. J Dairy Sci 99:5681–5689. https://doi.org/10.3168/jds.2015-10530

CONAFE (2024) Confederación de Asociaciones de Frisona Española. https://www.conafe.com/Home.aspx. Accessed 18 Jan 2024

Davis H, Stergiadis S, Chatzidimitriou E, Sanderson R, Leifert C, Butler G (2020) Meeting breeding potential in organic and low-input dairy farming. Front Vet Sci 7:1–13. https://doi.org/10.3389/fvets.2020.544149

de Araújo JV, Guimarães MP, Campos AK, de Sá NC, Sarti P, Assis RCL (2004) Control of bovine gastrointestinal nematode parasites using pellets of the nematode-trapping fungus Monacrosporium thaumasium. Ciência Rural 34:457–463. https://doi.org/10.1590/s0103-84782004000200019

de Haas Y, Smolders EAA, Hoorneman JN, Nauta WJ, Veerkamp RF (2013) Suitability of cross-bred cows for organic farms based on cross-breeding effects on production and functional traits. Animal 7:655–664. https://doi.org/10.1017/s1751731112002042

Dechow CD, Rogers GW, Cooper JB, Phelps MI, Mosholder AL (2007) Milk, fat, protein, somatic cell score, and days open among Holstein, Brown Swiss, and their crosses. J Dairy Sci 90:3542–3549. https://doi.org/10.3168/jds.2006-889

Dhakal K, Maltecca C, Cassady J, Baloche G, Williams C, Washburn SP (2013) Calf birth weight, gestation length, calving ease, and neonatal calf mortality in Holstein, Jersey, and crossbred cows in a pasture system. J Dairy Sci 96:690–698. https://doi.org/10.3168/jds.2012-5817

Dias AS, Araújo JV, Braga FR, Puppin AC, Perboni WR (2013) Pochonia chlamydosporia in the biological control of Fasciola hepatica in cattle in southeastern Brazil. Parasitol Res 112:2131–2136. https://doi.org/10.1007/s00436-013-3372-9

Eckard RJ, Grainger C, de Klein CAM (2010) Options for the abatement of methane and nitrous oxide from ruminant production: a review. Livest Sci 130:47–56. https://doi.org/10.1016/j.livsci.2010.02.010

EFSA (European Food Safety Authority) (2009) Effects of farming systems on dairy cow welfare and disease 1. Report of the Panel on Animal Health and Welfare. EFSA J 1143:1–284

Endendijk D, Baars T (2001) Family breeding at the Rivelinohoeve, a case study. In: Hovi M, Baars T (eds) 4th NAHWOA workshop.Network for Animal Health and Welfare in Organic Agriculture. University of Reading (UK), Wageningen, pp 35–43

Erlt P, Knaus W, Steingwidder A (2014) Comparison of zero concentrate supplementation with different quantities of concentrates in terms of production, animal health, and profitability of organic dairy farms in Austria. Org Agric 4:233–242. https://doi.org/10.1007/s13165-014-0077-z

Erskine RJ, Eberhart RJ (1991) Post-milking teat dip use in dairy herds with high or low somatic cell counts. J Am Vet Med Assoc 199:1734–1736

European Commission (2007) Council Regulation (EC) N° 834/2007 of 28 June 2007 on organic production and labelling of organic products and repealling Regulation (ECC) N° 2092/91. Off J Eur Union L189:1–23

European Commission (2018) Regulation (EU) 2018/848 on organic production and labelling of organic product. Off J Eur Union 2018:1–92

European Commission (2021) EU 2021/1165. Commission Implementing Regulation (EU) 2021/1165 of 15 July 2021 authorising certain products and substances for use in organic production and establishing their lists.C/2021/5149. Off J Eur Union 253:13–48

European Commission (2023) Organic farming in the EU – a decade of organic growth. Organic farming in the EU: a decade of growth, p 32

European Council (2023) Infographic – how EU countries are addressing the global food crisis. https://www.consilium.europa.eu/en/infographics/how-eu-countries-are-addressing-the-global-food-crisis/. Accessed 28 Jan 2024
European Parliament (2009) Directive 2009/128/EC of the European Parliament and the Council of 21 October 2009 establishing a framework for Community action to achieve the sustainable use of pesticides. Off J Eur Union 309:71–86
EUROSTAT (2023) Developments in organic farming. https://ec.europa.eu/eurostat/statistics-explained/index.php?title=Developments_in_organic_farming&oldid=614575#Organic_production. Accessed 28 Jan 2024
Fernandes L, Guimaraes I, Noyes NR, Caixeta LS, Machado VS (2021) Effect of subclinical mastitis detected in the first month of lactation on somatic cell count linear scores, milk yield, fertility, and culling of dairy cows in certified organic herds. J Dairy Sci 104:2140–2150. https://doi.org/10.3168/jds.2020-19153
Ferreras MC, González-Lanza C, Pérez V, Fuertes M, Benavides J, Mezo M, González-Warleta M, Giráldez J, Martínez-Ibeas AM, Delgado L, Fernández M, Manga-González MY (2014) Calicophoron daubneyi (Paramphistomidae) in slaughtered cattle in Castilla y León (Spain). Vet Parasitol 199:268–271. https://doi.org/10.1016/j.vetpar.2013.10.019
Flachowsky G, Franke K, Meyer U, Leiterer M, Schöne F (2014) Influencing factors on iodine content of cow milk. Eur J Nutr 53:351–365. https://doi.org/10.1007/s00394-013-0597-4
Forbes AB, Vercruysse J, Charlier J (2008) A survey of the exposure to Ostertagia ostertagi in dairy cow herds in Europe through the measurement of antibodies in milk samples from the bulk tank. Vet Parasitol 157:100–107. https://doi.org/10.1016/j.vetpar.2008.06.023
Fouz R, Yus E, Sanjuán ML, Diéguez FJ (2014) Causas de eliminación en rebaños bovinos lecheros de raza frisona en Control Lechero Oficial. Inf Tec Eco Agrar 110:171–186
González-Warleta M, Lladosa S, Castro-Hermida JA, Martínez-Ibeas AM, Conesa D, Muñoz F, López-Quílez A, Manga-González Y, Mezo M (2013) Bovine paramphistomosis in Galicia (Spain): prevalence, intensity, aetiology and geospatial distribution of the infection. Vet Parasitol 191:252–263. https://doi.org/10.1016/j.vetpar.2012.09.006
Green M, Bradley A, Newton H, Browne WJ (2006) Seasonal variation of bulk milk somatic cell counts in UK dairy herds: investigations of the summer rise. Prev Vet Med 74:293–308. https://doi.org/10.1016/j.prevetmed.2005.12.005
Haas E, Bapst B (2004) Swiss organic dairy farmer survey: which path for the organic cow in the future? In: 2nd SAFO Workshop Organic livestock farming: potential and limitations of husbandry practice to secure animal health and welfare and food quality. Witzenhausen (Germany), pp 35–41
Hamilton C, Emanuelson U, Forslund K, Hansson I, Ekman T (2006) Mastitis and related management factors in certified organic dairy herds in Sweden. Acta Vet Scand 48:11. https://doi.org/10.1186/1751-0147-48-11
Hammon HM, Stürmer G, Schneider F, Tuchscherer A, Blum H, Engelhard T, Genzel A, Staufenbiel R, Kanitz W (2009) Performance and metabolic and endocrine changes with emphasis on glucose metabolism in high-yielding dairy cows with high and low fat content in liver after calving. J Dairy Sci 92:1554–1566. https://doi.org/10.3168/jds.2008-1634
Hansen SL, Spears JW (2009) Bioaccessibility of iron from soil is increased by silage fermentation. J Dairy Sci 92:2896–2905. https://doi.org/10.3168/jds.2008-1933
Hardie CA, Wattiaux M, Dutreuil M, Gildersleeve R, Keuler NS, Cabrera VE (2014) Feeding strategies on certified organic dairy farms in Wisconsin and their effect on milk production and income over feed costs. J Dairy Sci 97:4612–4623. https://doi.org/10.3168/jds.2013-7763
Heins BJ, Hansen LB, Seykora AJ (2006) Production of pure Holsteins versus crossbreds of Holstein with Normande, Montbeliarde, and Scandinavian Red. J Dairy Sci 89:2799–2804. https://doi.org/10.3168/jds.S0022-0302(06)72356-6
Hofstetter P, Frey HJ, Gazzarin C, Wyss U, Kunz P (2014) Dairy farming: indoor v. pasture-based feeding. J Agric Sci 152:994–1011. https://doi.org/10.1017/S0021859614000227

Hogan JS, Smith KL, Hoblet KH, Schoenberger PS, Todhunter DA, Hueston WD, Pritchard DE, Bowman GL, Heider LE, Brockett BL, Conrad HR (1989) Field survey of clinical mastitis in low somatic cell count herds. J Dairy Sci 72:1547–1556. https://doi.org/10.3168/jds.S0022-0302(89)79266-3

Höglund J, Dahlström F, Engström A, Hessle A, Jakubek EB, Schnieder T, Strube C, Sollenberg S (2010) Antibodies to major pasture borne helminth infections in bulk-tank milk samples from organic and nearby conventional dairy herds in south-central Sweden. Vet Parasitol 171:293–299. https://doi.org/10.1016/j.vetpar.2010.04.002

Horn M, Steinwidder A, Podstatzky L, Gasteiner J, Zollitsch W (2012) Comparison of two different dairy cow types in an organic, low input milk production system under Alpine conditions. Agric For Res 362:322–325

Ivemeyer S, Brinkmann J, March S, Simantke C, Winckler C, Knierim U (2018) Major organic dairy farm types in Germany and their farm, herd, and management characteristics. Org Agric 8:231–247. https://doi.org/10.1007/s13165-017-0189-3

Kabata-Pendias A (2011) Trace elements in soils and plants, 4th edn. CRC Press, Boca Raton

Keatinge R, Jackson F, Kyriazakis I, Cork S (2002) Developing parasite control strategies in organic systems. In: Proceedings of the UK Organic research 2002 conference, powell. Organic Centre Wales, Institute of Rural Studies, University of Wales Aberystwyth, pp 341–346

Koesling M, Hansen S, Bleken MA (2017) Variations in nitrogen utilisation on conventional and organic dairy farms in Norway. Agric Syst 157:11–21. https://doi.org/10.1016/j.agsy.2017.06.001

Lean I, Golder HM (2023) Pasture minerals for dairy cattle. Vet Clin North Am Food Anim Pract 39:439–458. https://doi.org/10.1016/j.cvfa.2023.05.003

Leiber F, Schenk IK, Maeschli A, Ivemeyer S, Zeitz JO, Moakes S, Klocke P, Staehli P, Notz C, Walkenhorst M (2017) Implications of feed concentrate reduction in organic grassland-based dairy systems: a long-term on-farm study. Animal 11:2051–2060. https://doi.org/10.1017/S1751731117000830

LIC (Livestock Improvement Corporation) (2024) https://www.lic.co.nz/. Accessed 17 Jan 2024

Linehan K, Patangia DV, Ross RP (2024) Production, composition and nutritional properties of organic milk: a critical review. Food Secur 13(4):1–23. https://doi.org/10.3390/foods13040550

Lombard J, Garry F, Tomlinson S, Garber L (2007) Impacts of dystocia on health and survival of dairy calves. J Dairy Sci 90:1751–1760. https://doi.org/10.3168/jds.2006-295

López-Alonso M (2012) Trace minerals and livestock: not too much not too little. ISRN Vet Sci 2012:1–18. https://doi.org/10.5402/2012/704825

Mancuso WA (2017) Evaluación y comparación de grupos genéticos lecheros en un sistema a pastoreo de la comarca lechera de Entre Ríos, Argentina. Universidade de Santiago de Compostela

Marković J, Štrbanović R, Cvetković M, Anđelković B, Živković B (2009) Effects of growth stage on the mineral concentrations in alfalfa (Medicago sativa L.) leaf, stem and the whole plant. Biotechnol Anim Husb 25:1225–1231

Mazumder A, Kumagai H (2006) Analyses of factors affecting dry matter intake of lactating dairy cows. Anim Sci J 77:53–62. https://doi.org/10.1111/j.1740-0929.2006.00320.x

McBride BW, Greene C (2009) Characteristics, costs and issues for organic dairy farming. USDA-ERS report no. 82. https://www.ers.usda.gov/webdocs/publications/46267/11004_err82_1_.pdf?v=41079%3E#:~:text=milkin2005.-ChallengesofOrganicMilkProduction.,production%2Candmaintaininganimalhealth. Accessed 22 Feb 2024

Mendoza-de Gives P, López-Arellano ME, Aguilar-Marcelino L, Olazarán-Jenkins S, Reyes-Guerrero D, Ramírez-Várgas G, Vega-Murillo VE (2018) The nematophagous fungus Duddingtonia flagrans reduces the gastrointestinal parasitic nematode larvae population in faeces of orally treated calves maintained under tropical conditions—Dose/response assessment. Vet Parasitol 263:66–72. https://doi.org/10.1016/j.vetpar.2018.10.001

Mezo M, González-Warleta M, Castro-Hermida JA, Ubeira FM (2008) Evaluation of the flukicide treatment policy for dairy cattle in Galicia (NW Spain). Vet Parasitol 157:235–243. https://doi.org/10.1016/j.vetpar.2008.07.032

Mezo M, González-Warleta M, Castro-Hermida JA, Muiño L, Ubeira FM (2011) Association between anti-F. hepatica antibody levels in milk and production losses in dairy cows. Vet Parasitol 180:237–242. https://doi.org/10.1016/j.vetpar.2011.03.009

Miranda M, López-Alonso M, Castillo C, Hernández J, Benedito JL (2005) Effects of moderate pollution on toxic and trace metal levels in calves from a polluted area of northern Spain. Environ Int 31:543–548. https://doi.org/10.1016/j.envint.2004.09.025

Mullen K, Dingst E, Kearns R, Washburn SP (2015) A comparison of production, reproduction, and animal health for pastured dairy cows managed either conventionally or with use of organic principles. Prof. Anim Sci 31:167–174. https://doi.org/10.15232/pas.2014-01367

Nauta WJ (2001) Breeding strategies for organic animal production, an international discussion. In: Hovi M, Baars T (eds) 4th NAHWOA workshop. Network for Animal Health and Welfare in Organic Agriculture. University of Reading (UK), Wageningen (The Netherlands), pp 4–13

Nauta WJ, Veerkamp RF, Brascamp EW, Bovenhuis H (2006) Genotype by environment interaction for milk production traits between organic and conventional dairy cattle production in The Netherlands. J Dairy Sci 89:2729–2737. https://doi.org/10.3168/jds.S0022-0302(06)72349-9

Nauta WJ, Baars T, Saatkamp H, Weenink D, Roep D (2009) Farming strategies in organic dairy farming: effects on breeding goal and choice of breed. An explorative study. Livest Sci 121:187–199. https://doi.org/10.1016/j.livsci.2008.06.011

Nielsen M, Schukken Y, Scholl D, Wilbrink H, Brand A (1989) Twinning in dairy cattle: a study of risk factors and effects. Theriogenology 32:845–862. https://doi.org/10.1016/0093-691x(89)90473-1

Orjales I, López-Alonso M, Rodríguez-Bermúdez R, Rey-Crespo F, Villar A, Miranda M (2016) Is lack of antibiotic usage affecting udder health status of organic dairy cattle? J Dairy Res 83:464–467. https://doi.org/10.1017/S0022029916000601

Orjales I, Lopez-Alonso M, Miranda M, Rodríguez-Bermúdez R, Rey-Crespo F, Villar A (2017) The main factors affecting somatic cell count in organic dairy farming. Span J Agric Res 15. https://doi.org/10.5424/sjar/2017154-11769

Orjales I, Herrero-Latorre C, Miranda M, Rey-Crespo F, Rodríguez-Bermúdez R, López-Alonso M (2018) Evaluation of trace element status of organic dairy cattle. Animal 12:1296–1305. https://doi.org/10.1017/S1751731117002890

Orjales I, Lopez-Alonso M, Miranda M, Alaiz-Moretón H, Resch C, López S (2019) Dairy cow nutrition in organic farming systems. Comparison with the conventional system. Animal 13:1084–1093. https://doi.org/10.1017/S1751731118002392

Overton TR, Yasui T (2014) Practical applications of trace minerals for dairy cattle. J Anim Sci 92:416–426. https://doi.org/10.2527/jas.2013-7145

Pablos-Tanarro A, Pérez-Cabal MÁ, Ortega-Mora LM, Ferre I (2013) Presence of Ostertagia ostertagi antibodies in bulk tank milk from cattle herds in northern Spain. Vet Parasitol 197:388–392. https://doi.org/10.1016/j.vetpar.2013.05.021

Peeters A, Wezel A (2017) Agroecological principles and practices for grassbased farming systems. In: Wezel A (ed) Agroecological practices for sustainable agriculture: principles, applications, and making the transition. World Scientific, New Jersey, pp 293–354

Peña-Mosca F, Dean C, Machado V, Fernandes L, Pinedo P, Doster E, Heins B, Sharpe K, Ray T, Feijoo V, Antunes A, Baumann C, Wehri T, Noyes N, Caixeta L (2023a) Investigation of intramammary infections in primiparous cows during early lactation on organic dairy farms. J Dairy Sci 106:9377–9392. https://doi.org/10.3168/jds.2022-23036

Peña-Mosca F, Dean C, Fernandes L, Doster E, Sharpe K, Ray T, Feijoo V, Antunes A, Baumann C, Wehri T, Heins B, Pinedo P, Machado V, Noyes N, Caixeta L (2023b) Associations between early lactation intramammary infections and udder health and performance during the first 180 days in milk in first-lactation organic dairy cows. J Dairy Sci 107:2426–2443. https://doi.org/10.3168/jds.2023-23924

Ploeger HW, Verbeek PC, Dekkers CMH, Strube C, Van Engelen E, Uiterwijk M, Lam TJGM, Holzhauer M (2012) The value of a bulk-tank milk ELISA and individual serological and faecal examination for diagnosing (sub)clinical Dictyocaulus viviparus infection in dairy cows. Vet Parasitol 184:168–179. https://doi.org/10.1016/j.vetpar.2011.08.032

Randby ÅT, Weisbjerg MR, Nørgaard P, Heringstad B (2012) Early lactation feed intake and milk yield responses of dairy cows offered grass silages harvested at early maturity stages. J Dairy Sci 95:304–317. https://doi.org/10.3168/jds.2011-4454

Renau J (1986) Effective use of dairy herd improvement somatic cell counts in mastitis control. J Dairy Sci 69:1708–1720. https://doi.org/10.3168/jds.S0022-0302(86)80590-2

Rengab-Genetics (2013) Dairy meat breeds. Performance dual purpose breeds for crossbreeding. http://www.rengab-dairymeat.info/dairy-meat-breeds%2D%2D-3.html. Accessed 14 Dec 2016

Rey-Crespo F, Miranda M, López-Alonso M (2013) Essential trace and toxic element concentrations in organic and conventional milk in NW Spain. Food Chem Toxicol 55:513–518. https://doi.org/10.1016/j.fct.2013.01.040

Rey-Crespo F, López-Alonso M, Miranda M (2014) The use of seaweed from the Galician coast as a mineral supplement in organic dairy cattle. Animal 8:580–586. https://doi.org/10.1017/S1751731113002474

Roche JR, Berry DP, Delaby L, Dillon PG, Horan B, Macdonald KA, Neal M (2018) Review: new considerations to refine breeding objectives of dairy cows for increasing robustness and sustainability of grass-based milk production systems. Animal 12:S350–S362. https://doi.org/10.1017/S1751731118002471

Rodríguez-Bermúdez R, López-Alonso M, Rey-Crespo F, Cortés L, Orjales I, Miranda M (2016) Raças de bovinos de leite em modo de produção biológico no norte de Espanha. In: Marta-Costa AA, Tibério ML, Payan-Carreira R (eds) Raças autóctones no espaço ibérico: Um recurso sustentável. Universidade de Trás-os-Montes e Alto Douro, Vila Real, pp 25–30

Rodríguez-Bermúdez R, Miranda M, Orjales I, Rey-Crespo F, Muñoz N, López-Alonso M (2017) Holstein-friesian milk performance in organic farming in North Spain: comparison with other systems and breeds. Span J Agric Res 15:e0601. https://doi.org/10.5424/sjar/2017151-10037

Rodríguez-Bermúdez R, Fouz R, Miranda M, Orjales I, Minervino AHH, López-Alonso M (2019a) Organic or conventional dairy farming in northern Spain: Impacts on cow reproductive performance. Reprod Domest Anim 54. https://doi.org/10.1111/rda.13446

Rodríguez-Bermúdez R, Miranda M, Baudracco J, Fouz R, Pereira V, López-Alonso M (2019b) Breeding for organic dairy farming: what types of cows are needed? J Dairy Res 86:3–12. https://doi.org/10.1017/S0022029919000141

Rodríguez-Bermúdez R, Miranda M, Fouz R, Orjales I, Diéguez FJ, Minervino AHH, López-Alonso M (2020a) Breed performance in organic dairy farming in Northern Spain. Reprod Domest Anim 55:93–104. https://doi.org/10.1111/rda.13595

Rodríguez-Bermúdez R, Miranda M, Orjales I, Ginzo-Villamayor MJ, Al-Soufi W, López-Alonso M (2020b) Consumers' perception of and attitudes towards organic food in Galicia (Northern Spain). Int J Consum Stud 44:206–219. https://doi.org/10.1111/ijcs.12557

Roesch M, Doherr MG, Schären W, Schällibaum M, Blum JW (2007) Subclinical mastitis in dairy cows in Swiss organic and conventional production systems. J Dairy Res 74:86–92. https://doi.org/10.1017/S002202990600210X

Rose Vineer H, Morgan ER, Hertzberg H, Bartley DJ, Bosco A, Charlier J, Chartier C, Claerebout E, De Waal T, Hendrickx G, Hinney B, Höglund J, Jez Ek JI, Kašný M, Keane OM, Martínez-Valladares M, Mateus TL, McIntyre J, Mickiewicz M, Munoz AM, Phythian CJ, Ploeger HW, Rataj AV, Skuce PJ, Simin S, Sotiraki S, Spinu M, Stuen S, Thamsborg SM, Vadlejch J, Varady M, Von Samson-Himmelstjerna G, Rinaldi L (2020) Increasing importance of anthelmintic resistance in European livestock: creation and meta-analysis of an open database. Parasite 27:69. https://doi.org/10.1051/parasite/2020062

Rozzi P, Miglior F, Hand KJ (2007) A total merit selection index for Ontario organic dairy farmers. J Dairy Sci 90:1584–1593. https://doi.org/10.3168/jds.S0022-0302(07)71644-2

Ruegg PL (2017) A 100-year review: mastitis detection, management, and prevention. J Dairy Sci 100:10381–10397. https://doi.org/10.3168/jds.2017-13023

Rushen J, Passillé A (2013) The importance of improving cow longevity. In: Cow longevity conference, Tumba (Sweden), pp 3–21

Savory A, Bingham S, Butterfield J (2019) Holistic management. Regenerating your land and growing your profits. Island Press, Washington

Simm G, Villanueva B, Sinclair KD, Townsend S (2004) Farm animal genetic resources. Nottingham University Press, Nottingham

Slagboom M, Kargo M, Edwards D, Sørensen AC, Thomasen JR, Hjortø L (2016) Organic dairy farmers put more emphasis on production traits than conventional farmers. J Dairy Sci 99:9845–9856. https://doi.org/10.3168/jds.2016-11346

Slagboom M, Hjortø L, Sørensen AC, Mulder HA, Thomasen JR, Kargo M (2020) Possibilities for a specific breeding program for organic dairy production. J Dairy Sci 103:6332–6345. https://doi.org/10.3168/jds.2019-16900

Smith K, Hogan JS (1993) Environmental mastitis. Vet Clin North Am Food Anim Pract 9:489–498. https://doi.org/10.1016/S0749-0720(15)30616-2

Sorensen M, Norberg E, Pedersen J, Christensen L (2008) Invited review: crossbreeding in dairy cattle: a Danish perspective. J Dairy Sci 91:4116–4128. https://doi.org/10.3168/jds.2008-1273

Srednicka-Tober D, Barański M, Seal CJ, Sanderson R, Benbrook C, Steinshamn H, Gromadzka-Ostrowska J, Rembiałkowska E, Skwarło-Sońta K, Eyre M, Cozzi G, Larsen MK, Jordon T, Niggli U, Sakowski T, Calder PC, Burdge GC, Sotiraki S, Stefanakis A, Stergiadis S, Yolcu H, Chatzidimitriou E, Butler G, Stewart G, Leifert C (2016) Higher PUFA and n-3 PUFA, conjugated linoleic acid, α-tocopherol and iron, but lower iodine and selenium concentrations in organic milk: a systematic literature review and meta- and redundancy analyses. Br J Nutr 115:1043–1060. https://doi.org/10.1017/S0007114516000349

Sreenan JM, Diskin MG (1989) Effect of a unilateral or bilateral twin embryo distribution on twinning and embryo survival rate in the cow. J Reprod Fertil 87:657–664. https://doi.org/10.1530/jrf.0.0870657

Stiglbauer K, Cicconi-Hogan K, Richert R, Schukken Y, Ruegg PL, Gamroth M (2013) Assessment of herd management on organic and conventional dairy farms in the United States. J Dairy Sci 96:1290–1300. https://doi.org/10.3168/jds.2012-5845

Sundberg T, Berglund B, Rydhmer L, Strandberg E (2009) Fertility, somatic cell count and milk production in Swedish organic and conventional dairy herds. Livest Sci 126:176–182. https://doi.org/10.1016/j.livsci.2009.06.022

Sveberg G, Rogers GW, Cooper J, Refsdal AO, Erhard HW, Kommisrud E, Buckley F, Waldmann A, Ropstad E (2015) Comparison of Holstein-Friesian and Norwegian red dairy cattle for estrus length and estrous signs. J Dairy Sci 98:2450–2461. https://doi.org/10.3168/jds.2014-8905

Swalve HH (2007) Crossbreeding in dairy cattle: international trends and results from crossbreeding data in Germany. Lohmann Inf 42:38–46

Swecker WS (2023) Trace mineral feeding and assessment. Vet Clin North Am Food Anim Pract 39:385–397. https://doi.org/10.1016/j.cvfa.2023.05.001

Swinkels J, Lam T, Green M, Bradley A (2013) Effect of extended cefquinome treatment on clinical persistence or recurrence of environmental clinical mastitis. Vet J 197:682–687. https://doi.org/10.1016/j.tvjl.2013.03.010

TEAGASC (2023) Organic dairy farming. Agriculture and food: development. https://www.teagasc.ie/media/website/rural-economy/rural-development/diversification/1-Organic-Dairy-Farming.pdf. Accessed 12 Jan 2024

Thamsborg SM, Roepstorff A, Nejsum P, Mejer H (2010) Alternative approaches to control of parasites in livestock: Nordic and Baltic perspectives. Acta Vet Scand 52:1–5. https://doi.org/10.1186/1751-0147-52-s1-s27

Thomsen P, Ostergaard S, Houe H, Sorensen J (2007) Loser cows in Danish Dairy herds. Prev Vet Med 79:136–154

Thornton I, Abrahams P (1983) Soil ingestion, major pathway of heavy metals into livestock grazing contaminated land. Sci Total Environ 28:287–294. https://doi.org/10.1016/S0048-9697(83)80026-6

Toolan DP, Mitchell G, Searle K, Sheehan M, Skuce PJ, Zadoks RN (2015) Bovine and ovine rumen fluke in Ireland-prevalence, risk factors and species identity based on passive veterinary surveillance and abattoir findings. Vet Parasitol 212:168–174. https://doi.org/10.1016/j.vetpar.2015.07.040

Undersander D, Albert B, Cosgrove D, Johnson D, Peterson P (2002) Pastures for profit: a guide to rotational grazing (A3529). https://learningstore.extension.wisc.edu/products/pastures-for-profit-a-guide-to-rotational-grazing-p96. Accessed 20 Apr 2018

Vaarst M, Bennedsgaard T, Klaas I, Nissen T, Thamsborg S, Ostergaard S (2006) Development and daily management of an explicit strategy of non use of antimicrobial drugs in twelve Danish organic dairy herds. J Dairy Sci 89:1842–1853. https://doi.org/10.3168/jds.S0022-0302(06)72253-6

Valckenier D, Piepers S, De Visscher A, De Vliegher S (2020) The effect of intramammary infection in early lactation with non-aureus staphylococci in general and Staphylococcus chromogenes specifically on quarter milk somatic cell count and quarter milk yield. J Dairy Sci 103:768–782. https://doi.org/10.3168/jds.2019-16818

Van Diepen P, Mclean B, Frost D (2007) Livestock breeds and organic farming systems. ADAS Pwllpeiran, pp 1–39

Van Wagenberg CPA, De Haas Y, Hogeveen H, Van Krimpen MM, Meuwissen MPM, Van Middelaar CE, Rodenburg TB (2017) Animal Board Invited Review: comparing conventional and organic livestock production systems on different aspects of sustainability. Animal 11:1839–1851. https://doi.org/10.1017/S175173111700115X

Villar A, López-Alonso M (2015) Udder health in organic dairy cattle in Northern Spain. Span J Agric Res 13:1–8. https://doi.org/10.5424/sjar/2015133-6610

Villar A, Gradillas G, Fernández-Ruiz C, Rodríguez-Bermúdez R, López-Alonso M (2016) Dynamics of mammary infections in organic dairy farms in northern Spain. Span J Agric Res 14:e0502. https://doi.org/10.5424/sjar/2016142-7626

Voinot M, Cazapal-Monteiro C, Hernández JÁ, Palomero AM, Arroyo FL, Sanchís J, Pedreira J, Sánchez-Andrade R, Paz-Silva A, Arias MS (2020) Integrating the control of helminths in dairy cattle: deworming, rotational grazing and nutritional pellets with parasiticide fungi. Vet Parasitol 278:109038. https://doi.org/10.1016/j.vetpar.2020.109038

Vollema AR, Groen AF (1996) Genetic parameters of longevity traits of an upgrading population of dairy cattle. J Dairy Sci 79:2261–2267. https://doi.org/10.3168/jds.S0022-0302(96)76603-1

Wagner K, Brinkmann J, Bergschmidt A, Renziehausen C, March S (2021) The effects of farming systems (organic vs. conventional) on dairy cow welfare, based on the Welfare Quality® protocol. Animal 15. https://doi.org/10.1016/j.animal.2021.100301

Weller RF, Bowling PJ (2007) The importance of nutrient balance, cropping strategy and quality of dairy cow diets in sustainable organic systems. J Sci Food Agric 87:2768–2773. https://doi.org/10.1002/jsfa.3001

Chapter 8
Lamb Production in Patagonia

Sergio Radic-Schilling, Francisco Sales, Raúl Lira, René Muñoz-Arriagada, Paulo Corti, Iván Ordoñez, Nilo Covacevich, Ricardo Echeverría, Camila Sandoval, and Jorge Ivelic-Sáez

Abstract The lamb production area in Chilean Patagonia are located to the east in the Andean rain shadow, on the continental part and Tierra del Fuego Island, occuping an area of 35,962.6 km^2. The lamb meet is a relevant market today in this region, but is important to improve the grassland and rangeland degradation to maintain the quality of this product for the future. For latter, to improve the lamb production, grazing managemnet is the key factor where the calculate of carryng capacity is the primary determiantion to sustainable farm production. But also another points are important, to underpin the description of lamb systems, different characteristics are detailed in this chapter; as the geographical area, weather, rangelands soils, Magellan grazing ecosystem, lambs production, health and diseases, lamb and wildlife coexistence, welfare and market perspectives.

Keywords Lamb production · Chilean Patagonia · Rangeland · Lamb grazing · Wildlife interaction

---

S. Radic-Schilling (✉)
University of Magallanes, Punta Arenas, Chile

Chilean Academy of Agricultural Sciences, Punta Arenas, Chile
e-mail: sergio.radic@umag.cl

F. Sales · R. Lira · I. Ordoñez · C. Sandoval · J. Ivelic-Sáez
Agricultural Research Institute, Kampenaike, Punta Arenas, Chile

R. Muñoz-Arriagada · R. Echeverría
University of Magallanes, Punta Arenas, Chile

P. Corti
University Austral of Chile, Valdivia, Chile

N. Covacevich
Chilean Academy of Agricultural Sciences, Punta Arenas, Chile

C. Palacios Riocerezo, M. Gonzalez Ronquillo (eds.), *Traditional Livestock Production*, https://doi.org/10.1007/978-3-031-88275-3_8

8.1 Introduction

The last 148 years of sheep production in the Chilean Patagonia has transformed the region into the most important sheep farming area of Chile by the meat and wool production. Originally focused on wool production, sheep farming has seen a shift towards lamb production, which now generates the majority of the income in farm. Low rainfall and climate restrictions in general, condition the growth of rangelands, which directly impacts sheep productivity. However, these same factors contribute to the absence of certain diseases and pests, allowing the region to maintain a high health status. Animal welfare and market prospects for Magellanic lamb, particularly in the organic sector, position lamb from this region with significant potential to stand out internationally. Improvements have been realized through a strategic use of management techniques, new technologies and product quality, forward to sustainability management.

To underpin the description of lamb systems, different characteristics are detailed in this chapter; as the geographical area, weather, rangelands soils, Magellan grazing ecosystem, lambs production, health and diseases, lamb and wildlife coexistence, welfare and market perspectives.

8.2 Geographical Area and Weather

Located in the southern section of the American continent and also in Chilean territory. The region of Magallanes and Chilean Antarctica corresponds to an extensive territory (132,033.5 km^2) made up of the south-western area of Patagonia, the western part of the big Island of Tierra del Fuego (binational island Chile—Argentina) and an archipelagic zone adjacent to the previous ones (Fig. 8.1). The lamb production area (LPA) is located on the eastern slope of Magallanes, which, since the beginning of livestock farming in this territory, around the year 1970, until today, occupies an area of 35,962.6 km^2 (Martinic 2006; Radic et al. 2022), representing 27.2% of the regional territory. The LPA is mainly concentrated in the provinces of Magallanes, Tierra del Fuego and Última Esperanza (43.3%, 37.5% and 18.2% respectively) (Fig. 8.1). Orographically, the LPA is mainly characterized by the presence of eastern plains, these being low-elevation lands, with flat to smooth reliefs. Part of them correspond to the southern extension of the Patagonian plateaus (Pisano 1977). Next to the west, the eastern subandean zone appears, characterized by mountain ranges of low altitudes (500 m above sea level), while, to a lesser extent and mainly in the province of Última Esperanza, some mountain ranges appear, representing the southernmost section of the Patagonian Andes (Pisano 1977). The geographical breadth of the LPA in Magallanes gives way to 2 predominant climate types. According to the Köppen climate classification: the cold steppe climate and the trans-Andean climate with steppe degeneration. The first, associated with the sectors of the eastern plains is characterized by having short, cool summers

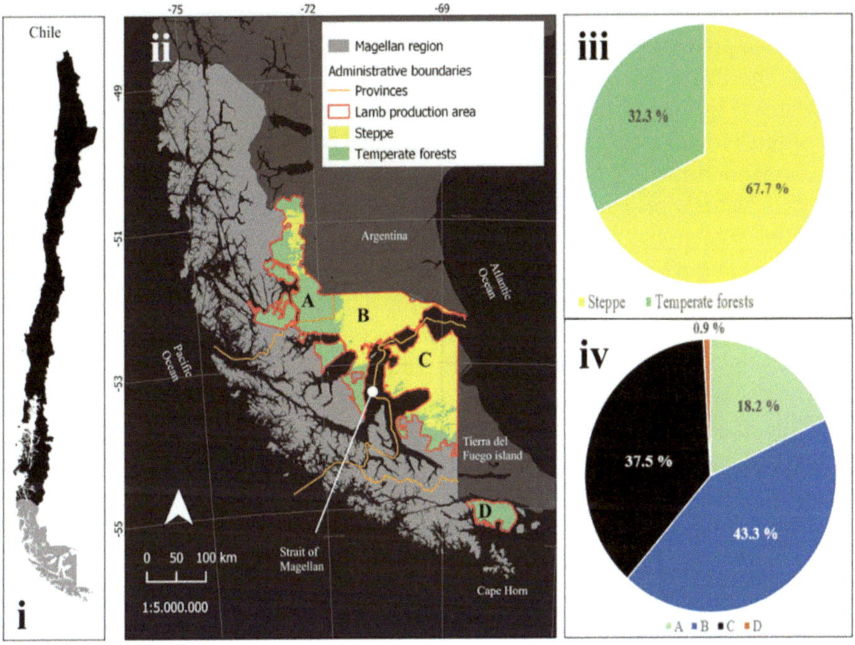

Fig. 8.1 (i) Shows the Chilean territory *in black* and the region of Magallanes and Chilean Antarctica *in gray*. (ii) Shows the Magellan lamb production area and the biomes present in it. The limits of the provinces are also shown: A = Última Esperanza, B = Magallanes, C = Tierra del Fuego and D = Chilean Antarctica. (iii) Shows a graph with the percentages according to biome for the lamb production area. (iv) Shows a graph with the percentages of lamb production areas by province. (Source: adapted from Radic et al. 2021, 2022)

and winters with averages ranging between 0 and 10 ° C (Xercavins 1984). The rainfall is low and fluctuates between 500 mm per year on the western slope and decreases towards the east, reaching sectors with less than 200 mm per year on the border with Argentine territory (Butorovic 2019). For its part, the trans-Andean climate with steppe degeneration, associated with the eastern sub-Andean region, is characterized by a precisely transitional behavior between rainy territories located to the west of the region and those associated with the steppe climate. Here rainfall fluctuates between just over 400 mm (on the border with the steppe) and 620 mm annually (in the westernmost section). While average annual temperatures are generally lower than 6.6 ° C (Pisano 1977). It should be noted that another important factor in the area corresponds to the wind, which has a predominantly western direction (westerlies), with average annual speeds that, in the LPA, can range from 4.6 m/s in the western area, and increase towards the east, exceeding 7.6 m/s (Endlicher and Santana 1988). These winds have a marked influence on the vegetation present in the area, generating mechanical and ecological effects. It is worth mentioning that potential evapotranspiration in the area can fluctuate between 480 and 700 mm year^{-1} (Radic et al. 2011).

The marked physical and climatic gradient of the region gives way to the presence of two predominant biomes for the LPA: the Patagonian steppe and temperate forests. The steppe represents 67.7% of the LPA, while the forests represent 23.2% of the area (Fig. 8.1). Temperate forests, which represent the southernmost distribution of this biome (Rozzi et al. 2012), are located on the western slope of the LPA and correspond mainly to deciduous Magellanic forest, where the characteristic species is *Nothofagus pumilio* (lenga), which can be associated in more humid sectors with *Nothofagus betuloides* (Magallanes coigüe) and in drier sectors with *Nothofagus antarctica* (ñirre) (Pisano 1977). In fact, in many cases (e.g. province of Magallanes) ñirre forests represent a transitional belt between deciduous forests and the steppe. It is worth mentioning that, due to historical forest fires (Otero 2006), especially in Magallanes and Última Esperanza, there is a strip where extensive areas of the so-called naturalized grasslands, originated from said fires, make way.

For another part, the Patagonian steppe occurs on the eastern slope of the provinces of Magallanes and Última Esperanza, while, in the province of Tierra del Fuego, it occupies the central and northern section of the Island. The predominant vegetation here corresponds to herbaceous and shrub communities, without the presence of trees. Within these we can mention: i) natural grasslands, where we distinguish: coironales or rangelands dominated by species of the Festuca genus (mainly *Festuca gracillima*), mesic grasslands or also known as "vegas", which represent the wetter areas of this biome and the saline grasslands. ii) heaths, mainly *Empetrum rubrum* (murtilla) and iii) scrublands, mainly *Chilliotrichium iffusum* (romerillo) and others with more limited distributions (Radic et al. 2021).

8.3 Rangelands Soils

The soils of the Magallanes region have mainly developed on the moraine systems of the second and third glaciation or on the mixture of these sediments. These soils contain abundant ashes from Pleistocene-Quaternary volcanism mixed with sand and fluvial and/or colluvial silt (Pisano 1977).

From east to west, the soils of the Magallanes region present an aridity that conditions the type of vegetation and soil, from semi-arid, less leached (Pisano 1977), with a higher concentration of cations, less acidic (Sáez 1994) and with steppe vegetation where the Tussock grass (*Festuca gracillima*) dominates (Radic et al. 2021), to soils with higher moisture content (Pisano 1977), lower concentration of cations, higher acidity (pH < 5.7) (Sáez 1994) and with shrub and tree vegetation (Pisano 1977; Radic et al. 2021). These soils have an upper profile with organic matter, while at depth, the textures change to clay loam and clay, being stony throughout the profile (Ordoñez et al. 2023) with the exception of the meadows. Furthermore, the Tussock grasses present in the steppe have an soil bulk density that varies between 0.58 and 0.87 g cm^{-3} (Radic et al. 2021) and a low content of nitrogen, phosphorus and sulfur (Ordoñez et al. 2023) while in shrub areas the soil bulk density varies between 0.82 and 1.44 g cm^{-3} (Radic et al. 2013).

The greatest vegetation coverage of the agricultural use area of the region is the steppe (Radic et al. 2021), where mollisols, inceptisols and aridisols soils predominate (USDA Taxonomy) (Ordoñez et al. 2023). Mollisol soils are formed under grassland vegetation in climates with moderate to marked water deficit, and are characterized by having a horizon where calcium is the dominant cation in the surface horizons (Luzio and Casanova 2006). Inceptisol soils are characterized by being incipient with slight morphological development in the subsoil (Hepp and Stolpe 2014). On the other hand, aridisol soils are characterized by not having water available when the temperature is adequate for the growth of plant species (Luzio and Casanova 2006). While in the more humid sectors, of bushes and forest, types of soil such as podzols (Spodosols—USDA taxonomy) can be found, and have higher levels of acidity and exchangeable aluminum (Sáez 1994).

Mainly in the steppes, there are other types of soils with azonal characteristics generated by humid depressions, or hydromorphic soils also called "vegas" (Valle et al. 2015; Ivelic-Sáez et al. 2021) and that present the highest soil fertility among the different plant communities of the region (Ordoñez et al. 2023). These humid depressions are mainly classified as Histosols, Entisols, and Inceptisols in which there is considerable variability in their organic matter content and acidity (Ordoñez et al. 2023). These meadow soils have a soil bulk density that varies between 0.20 and 0.39 g cm^{-3} (Radic et al. 2013) and the highest organic carbon with inclusive 30% (Ordoñez et al. 2023).

Although the soils of the region respond to fertilization with N, P and S (Ordoñez et al. 2023), the plant communities that generate the lowest primary productivity are those of murtillares and Tussock grass, while the communities of meadows and humid grasslands have the best performance (Navarro 2021).

8.4 Magellan Grazing Ecosystem

The pastoral ecosystems in Magallanes are divided by the sub-Antarctic rangelands and "vegas" (meadows) (Radic-Schilling et al. 2021). The rangelands are dominated by dense tussocks types of perennial species (i.e. *Festuca gracillima*, *Festuca pallescens* and *Stipa humulis*) and less unpalatable shrubs such as *Chiliotrichium diffusum* (Pissano 1973, 1977; Soto 1984). Coupled with an interpatch, composed by species from the family asteraceae, gramineae, cyperaceae among others (Soto 1984). The species that can be found in vegas of good condition are *Holcus lanatus*, *Juncus spp*, *Taraxacum officinalis*, *Carex spp*, *Hordeum spp*. among others (Filipová et al. 2013; Soto 1984). The presence of these types of species is because the water table is closer to the surface, delivering a greater offer of water than the rangeland (Ivelic-Sáez et al. 2021).

The rangelands showed an herbage mass potential around 800 kg DM ha^{-1} $year^{-1}$ (Covacevich 2001), with a variation during 2022, considering just *F. gracillima* growth, between 400 and 900 DM ha^{-1} $year^{-1}$, showing a greater growth rate between November–December (Ordóñez et al. 2023). The variation is explained by the different locations within the region. Furthermore, the crude protein varied was between 2.93%–4.13% and for metabolize energy between 1.06–1.46 Mcal kg^{-1} DM (Ordóñez et al. 2023) which does not fulfill the requirements of energy and protein for the sheep maintenance and growth (Strauch and Lira 2012). The vegas represent the 5% of the total productive surface in Magallan region (Lara and Cruz 1987) and an accumulated herbage mass between 500 kg DM to 11.000 DM ha^{-1} $year^{-1}$ (Covacevich 2001; Jara 2018; Ordóñez et al. 2023). The herbage mass production of a vega will depend on climate such as temperature and rainfall (Ordoñez et al. 2023), intrinsic properties (Filipová et al. 2010, 2013) and historical management (Collantes and Faggi 1999; Oliva et al. 2001). The crude proteins varied between 7.36%—18.89% and metabolize energy between 1.37–2.22 Mcal kg^{-1} DM (Ordóñez et al. 2023).

8.4.1 Magellan Grazing Systems

The grazing systems in Patagonia establish differential management criteria for the different vegetation structures (rangelands and vegas) and location within the region and landscape. Therefore, grazing criteria needs to be adapted to the plant traits together with its interaction with the environment and defoliation. The latter is because, (i) differences can be found related to the plant growth strategy (Reich 2014), such as species with a conservation of resources strategy (i.e. slow traits species such as *Festuca gracillima*) and acquisition of resources strategy (i.e. fast traits species such as *Holcus lanatus*) (Castro 2023) and (ii) a defoliated plant will respond differently depending on the conditions and climate where is located (for example rangeland or vega). Was observed that a defoliation intensity lower than 10 cm is detrimental for the growth of *F. gracillima* in simulated conditions (Oyanader, 2023). On the other hand, vegas dominated by high quality grasses can be utilized with a residual height of 5 cm because the grass energy reserves may be located between that height, as was observed in other grass species (Fulkerson and Donaghy 2001; Turner et al. 2006a, b, c, d). Furthermore, vegas are richer in resources such as water in comparison to the rangeland (Ivelic-Sáez et al. 2021; Ordóñez et al. 2023) making them more resilient and resistant to defoliation from the vegetation standpoint.

Independent of the grazing systems utilized, the separation of the different vegetations structures is important to control the grazing according to the herbage mass production and to avoid the animal selection, evading to overgraze the sites (Ormaechea et al. 2014) and soil compaction (Oliva 2001). Furthermore, are two relevant decisions for any grazing system; (i) establish a utilization criteria that ensure the vegetation regrowth, survival and soil health, this will be determined by the vegetation, soil type and climate; (ii) determine the correct stocking rate (Gammon 1978; Borrelli 2001) because determine the grazing intensity and frequency for the utilization of naturalized pasture, rangeland and vegas.

8.4.1.1 Seasonal Continuous Grazing System

Seasonal continuous grazing system is the most common in the extensive Magellan grazing management (Radic-Schilling et al. 2021). Is characterized by large paddocks designed to separate stock categories rather than to program defoliation periods (Radic-Shilling et al. 2021). The grazing is programmed by calendar (season), location of the grazing sites and reproductive cycles (Radic-Shilling et al. 2021). The grazing management splits in two seasons, summer grazing is performed between December to May in vegas or in sites over 150 m above sea level. Winter grazing is performed between May to December, in sites at sea level and/or closer to the sea due to a better temperature during winter (Covacevich 2012; Villa et al. 2020), and overlaps with prior lactation and postpartum periods (Romero and Bravo 2012). However, continuous grazing presents disadvantages that are the increasing heterogeneity (Borrelli 2001) and weeds invasion (Von Moltke 2009; Radic-Schilling et al. 2020).

8.4.1.2 Rotational Grazing

System that needs an improvement related to the farm organization, increasing the number of paddocks, refining the grazing in terms of the vegetation use (Ormaechea et al. 2014). However, the grazing time together with the animal movement must be calculated in such a way that does not affect the newborn lambs or the ewes in advanced pregnancy. The latter is critical due to the mortality of the sheep that can occur when the animal movements is over long distances. For this reason, rotational grazing has to be adapted to Magellan conditions (Fig. 8.2). Two types of rotational grazing can be applied, the rotational deferred grazing and traditional rotational grazing (Gammon 1978; Borrelli 2001; Covacevich 2012).

8.4.1.3 Traditional Rotational Grazing

This system is more feasible in high potential growth sites such as in locations with greater rainfalls or vegas. However, it is not recommendable to increase the animal density due to the diminishment in animal performance (Gammon 1978) and soil compaction (Marcos et al. 2020; Oliva 2001; Collantes and Faggi 1999). The vegetation in vegas of good condition that have a water table closer to the surface allow a fast recovery after the defoliation, however it needs to be considered not to overgraze the sites with highly intense defoliations (below 5 cm pasture height) or too frequent defoliations. It is relevant to allow the vegetation to recover to develop a sustainable grazing system (Fulkerson and Donghy 2001; García-Favre et al. 2023). Regarding to the rangeland, it is not recommendable the perform intensive grazing because *F. gracillima* show a slow growth and recovery (Ordóñez et al. 2022, 2023)

Fig. 8.2 Weanned lambs grazing in rotational system on a sward in Tierra del Fuego. (Photo by Sergio Radic)

in a highly constrained environment (Ivelic et al. 2021; Ordóñez et al. 2023). However, rotational grazing with lax criteria of utilization in rangelands may be an option to develop a sustainable grazing system. There is not enough evidence of the benefits of rotational grazing over continuous grazing in rangeland ecosystems under a proper stocking rate (Gammon 1978; Borrelli 2001; Briske et al. 2008).

8.4.1.4 Rotational Deferred Grazing

Is based on the rest of each paddock every specific number of years to allow its seeding (Gammon 1978) and its recovery after years of defoliation (Borrelli 2001). From the animal performance standpoint it was shown that there are no clear results when rotational deferred grazing is compared with continuous grazing (Gammon 1978). In fragile ecosystems such as in Magellan Region, may allow a recovery of the vegetation and soil properties due to the period without livestock. This type of grazing system may be beneficial for the rangeland, which is a highly restrictive ecosystem from the water standpoint and is dominated by a slow trait species (*F. gracillima*) with a slow recovery after defoliation. However, research has to be performed in relation to increase the resilience and the resistance of the ecosystems used for livestock production.

8.4.2 *Rangeland Degradation*

In the same line to the statements made by Briske et al. (2008) and Gammon (1978), the stocking rate is the more relevant decision in extensive grazing that shows a relevant impact on primary and secondary productivity. However, in continuous grazing and especially when the rotational grazing is applied, animal density can generate a degradation of the vegetation by overgrazing and soil compaction. The latter occurs when the number of animals is outside the range that does not allow an ecosystem to recover. The loss of soil functionality can be greater in soils with a low resilience and resistance to animal trampling, which may be the case for some vegetal communities such as tussock grass of *F. gracillima* and vegas, where the latter is the more resilient between both.

8.5 Lambs Production

Since the second half of the nineteenth century, the establishment and development of sheep ranching began in the southernmost region of Chile, Magallanes, Patagonia. Large land concessions by the State, and an important commitment from private companies allowed an accelerated growth of this industry (Lira 2012).

Natural rangelands in the world, which largely dominate the geography of Magallanes, are those areas used mainly for grazing because they cannot be cultivated. These include arid, and semiarid areas, that occupy 25% of Earth surface, with around 80% of beef, and sheep meat world production (Hacker 2010). In these large rangelands, the grazing system is "extensive" (Fig. 8.3), not only in the sense that it is carried out over large areas, but also because the level of inputs, and management of animals is relatively low.

In this typical Patagonian scenario, lamb meat was a by-product for many decades, with the sheep industry focused on wool production. That view changed in the last 30 to 40 years, driving the focus to meat production, which today explains most of the income, which turned dramatic from the world wool market collapse mainly observed from 2020 to 2021, at pandemia time.

Table 8.1, summarizes information comparing two different years, with official available data.

As observed in Table 8.1, over half of the sheep population is located in Magallanes, the southern Region of Chile. In this area, sheep production is the most important, and almost the only agricultural industry. It has been developed for the last 148 years, so it is part of the local culture, including a strong focus on wool production.

Main breed used to be Corriedale, with operations introducing different Merino streams (Australian Merino, Multi Purpose Merino, 4 M, and, most important in the latest years, Dohne Merino), and crosses during the last three decades. Because of the increasing value of meat in the total income of sheep business, it is common to

Fig. 8.3 Weanned lambs grazing on an improved pasture. (Photo by Raúl Lira)

Table 8.1 Stock (regional, and country), slaughtered animals, and carcass weight: 2007 and 2021[a]

Year:	2007	2021
Stock:		
Magallanes sheep stock (hd)	2,205,270	1,417,421
Chile sheep stock (hd)	3,888,485	2,529,238
Proportion Magallanes/Chile (%)	56.7	56.0
Magallanes: Number of slaughtered heads (number per year)		
All sheep	674,210	457,046
Yearlings (hoggets)	68,415	125,139
Lambs	512,434	251,804
Magallanes: Slaughtered carcass weight (kg):		
All sheep	13.8	15.3
Yearlings (hoggets)	21.0	17.6
Lambs	12.1	13.2

[a]National and Regional Statistics: Instituto Nacional de Estadísticas (INE), and Servicio Agrícola and Ganadero (SAG)

see terminal cross use, basically with Suffolk rams but also some Polled Dorset, Texel, and White Suffolk. Lamb crop is below 80%, as a regional average (Abella et al. 2010).

Again, based on Table 8.1, it is possible to observe a dramatic drop in both, national and regional stock comparing years 2021 and 2007, and proportionally, about the same; 35% for national, and local drop.

Highly probably, the reason why the drop in total processed number, and slaughter weight of yearlings is based in the increasing number of merinos in the local flock, as well as a degradation of rangeland. Same reason explains the increase in yearling animals coming into the slaughter-house, and the drop in the number of lambs.

Breeding season is in fall in Patagonia, and lambing in spring in an extensive management system with very low inputs, and low human intervention in results, which means erratic results in the meantime, basically depending on annual climatic conditions (Lira 2012).

Possible use of technologies in order to increase number, and weight of lambs coming into the slaughter-house, increasing income of the operation, can include: use of prolific breeds of ewes, or high specialization in meat production animals (hair sheep), terminal cross rams, strategic supplementation of dams, and replacements, pastures improvement and its planned use, scanning for gestation, and special management for ewes carrying twins, other practices.

8.6 Lamb Health and Diseases

Sheep production in Chilean Patagonia is developed mainly in extensive systems which are located in a region of challenging environmental conditions Among them, the low rainfall, cold weather, poor nutritional quality of the natural pasture which is the major nutrient source- are limitations for the success of an ovine operations (Sales et al. 2019). Particularly, the natural scenario of nutrient restriction given the quality of the pasture, and the extreme weather conditions, are the factors that would have a bigger impact on lamb mortality, diseases, and welfare. However, these conditions, and the natural isolation of the region are also major factors that limit the presence of disease agents, leading to a high sanitary status, similar to New Zealand (Radic et al. 2021), where chemicals are applied strategically, just to address the few specific pathologies that are present. This section will cover the main issues on lamb health and diseases that are present in the different periods of lamb production in Chilean Patagonia.

8.6.1 Lambing

Starving-Cold Exposure Syndrome The poor nutritional quality of the natural pasture in Chilean Patagonia leads to a natural undernutrition environment, which worsens during pregnancy when the nutrient requirements of the ewe are increased. This leads to low birthweight (i.e. 2.82 kg for twin lambs (Sales et al. 2019), and 4.72 kg for singles (Sandoval and Sales 2022) and reduced amounts of brown fat, the tissue used by the lamb for non-shivering thermogenesis. Hence, the neonates

Table 8.2 Major lamb diseases and prevalence in chilean patagonian systems

Lambing		Lambing to weaning		Finishing	
Disease	Prevalence	Disease	Prevalence	Disease	Prevalence
Starvation/cold	H	Coccidiosis	M	Enterotoxemia	M
Omphalitis	L	Gastrointestinal Parasites	L	Gastrointestinal Parasites	M
Arthritis	L	Lungworms	L	Lungworms	M
		External parasites	L	External parasites	M
		Pneumonia	L	Pink eye	L
		Diarrhea	L		

[a]H = High, M = Medium, L = Low

face difficulties for thermoregulation immediately after lambing, and given the extreme cold temperatures, this usually ends in hypothermia, starvation, and lamb death. This is a significant cause of lamb mortality, and one of the major factors that drops the tailing rate to an average of 75% in the region, representing a significant—if not, the most relevant-health, welfare, and economical problem of these systems.

Other low-prevalent health problems associated with this period can be seen in Table 8.2.

8.6.2 Lambing to Weaning

Coccidiosis This medium prevalence disease caused by *Eimeria spp.is* will produce diarrhea as a major sign, eventually impairing lamb growth potential, hence affecting the efficiency of the animal operation.

Ecthyma This zoonotic disease, caused by a Parapoxvirus (Tedla et al. 2018), is highly prevalent in this period as it is transmitted from lesions in the udder of the dam to the nursing lamb. It is characterized mainly by mouth lesions, but it can also sporadically lead to feet alterations. As it is a viral disease, its progression will be naturally limited so no treatment is usually required.

Parasites Dry, and cold weather conditions, low stocking rate, and summer-winter fields rotation practices are key aspects of Chilean Patagonian Systems that lead to a low gastrointestinal parasite burden. Even when round and tapeworms are normally present, the fecal egg counts of these parasites are low in lambs. Anthelmintic treatments are normally not applied to lambs before going to the slaughterhouse, with the exception of replacement hogget, which may receive anthelmintic at weaning. It is important to mention that, even when anthelmintics are not highly used, there are preliminary studies that have already found a risk for resistance against macrocyclic lactones (Sandoval and Sales 2022). This highlights the relevance of establishing well-organized control strategies that continue minimizing the use of chemical treatments as well as start considering the rotation of anthelmintic families. In regard to external parasites, sheep ked (*Melophagus ovinus*) and lice (*Damalinia ovis*) are present, and even when they could affect lambs, the problem is mostly confined to older animals.

Other diseases of low prevalence that may occur in this period can be seen in Table 8.2.

8.6.3 Finishing

Enterotoxemia In general, clostridial diseases are the most relevant pathologies encountered in Patagonian sheep farming. Among them, and specifically for finishing lambs, enterotoxemia is the most prevalent one. This disease, also known as pulpy kidney, is associated with explosive grass growth, sudden paddock changes or other managements involving feeding changes (West et al. 2009). The most characteristic clinical sign is diarrhea, fecal blood, and it could rapidly evolve to lamb death (Uzal and Songer 2008). Although the disease is confined to areas of higher humidity than Chilean Patagonia, it has been described throughout the region, affecting lambs at o after weaning time, which normally occurs at 4 months of age, resulting in animal losses. *Clostridium welchii*, the causing agent of pulpy kidney, in addition to *C. chauvoey* and *C. septicum*, are successfully controlled in lambs by maternal vaccination before lambing. As lambs are sold to market at an early stage of life, only replacement animals could get a prime vaccination at weaning, although it is not a common sanitary management.

Parasites As described in the previous section (lambing to weaning period), internal and external parasites are present in the region, but they do not usually reach critical levels in young or older animals. However, the prevalence may get slightly higher in lambs during this period, mostly if the sheep operation applies finishing strategies that involve the use of higher animal density during certain periods of time. However, data from our preliminary studies suggest that intensive grazing management during summer time, having a rotative system where the animals are not allowed to return into previously occupied areas, results in low parasite infections (unpublished data). Even with that, the use of anthelmintics remains low in lambs.

Other diseases such as pink eye, an eye bacterial infection (Egwu 1992), may occur in lambs during dry years. However, the prevalence is low.

8.7 Lamb Production and Wildlife Coexistence

The Patagonian steppe grasslands play a crucial role in both lamb production and the survival of diverse wildlife species. However, the extensive management of sheep production leads to interactions with wildlife, adversely affecting newborn sheep survival and, consequently, farmers' revenue (Radic et al. 2022). Addressing this issue often involves marginalization of wildlife from specific habitats, intensifying already conflictive relations.

Carnivorous native species, such as puma (*Puma concolor*), culpeo (*Lycalopex culapeus*), and grey fox (*L. griseous*) (Fig. 8.4), as well as exotics like the American mink (*Neogale vison*), pose a direct threat to lamb survival through predation (Bellati and Thungen, 1990). Another growing concern in Magallanes is predation on lambs by domestic dogs, aggravated by the lack of legal management regulation (Soto 2019). While lamb mortality due to predation in Argentinean Patagonia can reach ca. 45% (Bellati and Thungen 1990), several environmental variables, including heat loss, contribute to the remaining percentage (Coronato 1999). Although specific records on lamb predation rates in Magallanes are unavailable, domestic dogs (ca. 78%), both fox species (ca. 9%), and puma (ca. 1%) are the main causes of sheep losses (Soto 2019). Other potential lamb predators include raptors like the southern caracara (*Caracara plancus*) and buzzard eagle (*Geranoaetus melanoleucus*) (Bellati and Thungen 1990), although scientific information in Magallanes is lacking (Radic et al. 2022).

The guanaco (*Lama guanicoe*) (Fig. 8.4), a large native ungulate herbivore, is viewed as competitors for space and food with domestic sheep, potentially indirectly affecting lamb survival through competition for ewes' resources (González et al. 2022; Lancaster et al. 2022). Despite a significant overlap in the dietary composition of both species (Pontigo et al. 2020), guanaco food consumption is inferior (Lancaster et al. 2022), maybe due to the lower metabolism that characterize camelids (Dittmann et al. 2014; Lancaster et al. 2022). Recent research suggests that high sheep density, at times exceeding carrying capacity, adversely affects sheep feeding performance (Fernández et al.2021), thereby indirectly impacting newborn survival (Bonefant et al. 2009). Herbivorous birds sharing space with sheep, such as the lesser rhea (*Rhea pennata*), and wild geese (*Chlöephaga spp.*), are not considered competitors like guanaco (Radic et al. 2022), but Magallanes Livestock Farmers Associations are concern about new swards or reseedings and wild geese impact on it.

It is important to differentiate lamb losses caused by wildlife from those resulting from deficient ranch management (Bellati and Thungen 1990; Coronato 1999; Fernández et al. 2021; Lancaster et al. 2022). Several sheep farms appear to have inaccurate estimations of carrying capacity, leading to overgrazing and steppe degradation, ultimately resulting in poor ewe conditions and lamb losses unrelated to wild herbivores competition (Fernández et al. 2021). Before intending lethal control of carnivores and raptors, it is important to recognize their importance in maintaining ecosystem health. Additionally, considering improvements in sheep management practices (i.e., guard dogs; Ohrens et al. 2021), it is relevant to avoid predation on lambs.

Fig. 8.4 Grey fox (*L. griseous*) in the snow and guanaco (*Lama guanicoe*) grazing. (Photo by Paulo Corti)

8.8 Lamb Welfare

In the Magallanes Region, livestock is exploited mainly on the properties of medium and large producers on an extensive basis, registering an average herd of 4500 heads (INE 2017). However, a growing reduction in the number of producers has been observed over time, but an increase in the size of herds indicative of the concentration of property and an economy focused on farms with entrepreneurial characteristics.

In recent years, there has been great interest, concern not only for the livestock product, but for the way in which it is generated. The concept of "one health" has become relevant, in which animal health and the health of human beings Humans, interdependent with ecosystems, adopt a fundamental participation in sustainable development (De Queiroz et al. 2018).

The OIE defined animal welfare as those conditions in which "an animal is healthy, comfortable, well-fed, safe, in which it can express innate forms of behavior and does not suffer unpleasant sensations of pain or fear". This definition can be complemented with a list (Dwyer et al. 2008), which establishes that an animal is in conditions of well-being if it can experience the 5 freedoms: absence of hunger, thirst and malnutrition; absence of fear and sustained stress; absence of discomfort (physical and thermal, among others); absence of pain, injury and/or illness; and freedom to express natural behavior (SENASA 2015).

In search of a perspective on the application of good livestock practices and animal welfare in the Magallanes Region during the year 2023, an online survey was applied to farms that are clearly lamb producers, based on the 5 fundamental pillars of animal welfare. Preliminarily, out of a total of 46 respondents, 82.6% performed castration procedures and 100% performed tail docking and branding as identification, citing various health, reproductive and productive reasons, among others. However, various studies show us that when performing these procedures, the animal presents signs indicating pain such as changes in facial expression and behavior and even alterations in eye temperature. This last non-invasive technique allows detecting the infrared radiation emitted by the ocular surface, capturing the temperature changes associated with vasoconstriction generated by the release of catecholamines once castration bands are placed or the application of procedures such as branding or identification (Fig. 8.5).

Having pain indicators in sheep will contribute to its management and will make producers more aware of its importance as a welfare problem (Manteca et al. 2017). Indeed, guaranteeing the well-being of animals is important not only for ethical reasons, but also because many welfare problems at the same time are productive or health problems, converted today into requirements of demanding markets (Manteca 2012).

8.9 Lamb Market Perspectives

Chile ranks eighth among the top frozen lamb exporting countries with 0.74% of global export value or USDA $ 27.07 million (Verified Market Reports 2022) most of which is produced in Magallanes. Nevertheless, despite the steady increase in market trends in past years, prices have been variable, and production is decreasing because of climate change. Therefore, local producers received as good news the dietary preferences for organic meat as a healthier choice that have been observed in the last decades in developed countries. The lamb production extensive pastoral system in Magallanes evolved under natural conditions that comply with the requirements for organic lamb production offering animal products with unique origin features. Accordingly, by the year 2002, the Swiss company IMO Control certified

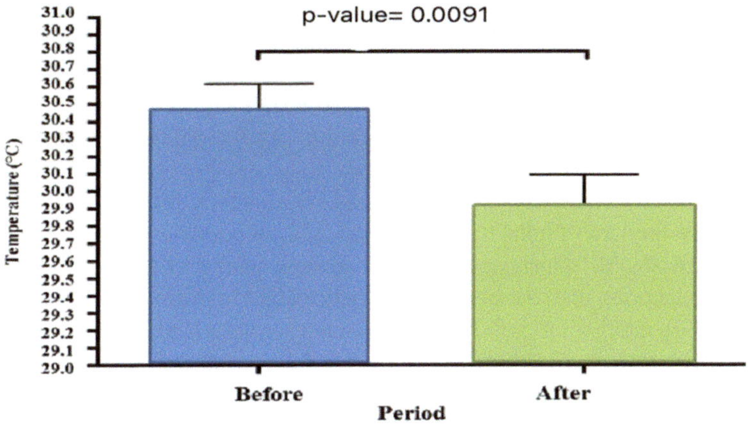

Fig. 8.5 Ocular temperature after elastic band placement. (Unpublished data Echeverría et al.)

265,711 ha as suitable for organic meat production in Magallanes. However, despite the first success in marketing tests, this certification did not increase revenue discouraging further efforts. At the time innovative technology preserving fresh processed meat ("chilled") became far more popular than its frozen raw counterpart which still is the traditional presentation of Magellan lamb products. Nowadays, however, both markets show positive trends. Despite the negative effects of the COVID-19 pandemic and the Russia-Ukraine war the forecast for the global Organic Lamb Market Size keeps growing steadily. It was estimated at USD 11720 million in 2022 and is projected to reach USD 17490 million by 2028, showing a CAGR of 6.9 (Precision Reports Market Growth Updates, https://www.linkedin.com/company/market-growth-updates/).

At present, local entrepreneurs backed by the certification of Global Animal Partnership (GAP) obtained for the year 2022 contracts to sell about seven hundred tons of organic fresh lamb meat in the US market. High-quality Magellan Organic Lamb, if complemented with modern processing and transport, should ensure significant participation in a global market where climate change and related product reduction is one of the major concerns of this industry.

8.10 Final Comments

The Magallanes Region in Chile has become a characteristic and important territory for lamb production, and meet is a relevant market today, but is important to improve the grassland and rangeland degradation to maintain the quality of this product for the future. In the same line, the stocking rate, rather than grazing systems, is the more relevant decision making in grazing management, for the latter it is important to make accurate calculation to determine the carrying capacity, because the lack of these calculations is the primary reason for rangeland degradation by overgrazing. For another side, the isolation and environmental conditions of Chilean Patagonia and the management of extensive systems, lead to exceptionally high sanitary standards in this region. This allows a good animal health and welfare and it is visualized as an appealing opportunity to advance into organic production certifications, at least from the animal health point of view.

To make a sustainable use of rangeland and grasslands, will support farm success and improve the lamb production. The latter will be possible through technology adoption, sustainable management and broadening market potential of lamb (i.e., organic production), also the best sheep practices that improve animal welfare and sheep-wildlife interactions will be useful to make a robust lamb production system.

Acknowledgments The authors acknowledge all technicians, students and scientists were involved in each research developed in the region in the last years and the farmers what contribute to the progress and sustainability of lamb production in Magallanes.

References

Abella I, Cardellino RC, Mueller J, Cardellino RA, Benítez D, Lira R (2010) South American sheep and wool industries. In: Cottle DJ (ed) International sheep and wool handbook. Nottingham University Press, Nottingham, pp 85–94

Bellati J, von Thungen J (1990) Lamb predation in Patagonian ranches. In: Davis LR, Marsh RE (eds) Proceedings 14th Vertebrate Pest Conference. University of California, Davis, pp 163–268

Bonenfant C, Gaillard JM, Coulson T, Festa-Bianchet M, Loison A, Garel M, Loe LE, Blanchard P, Pettorelli N, Owen-Smith N et al (2009) Empirical evidence of density-dependence in populations of large herbivores. In: Caswell H (ed) Advances in ecological research, vol 41. Academic Press, Burlington, pp 313–357

Borrelli P (2001) Producción animal sobre pastizales naturales. In: Borrelli P, Oliva G (eds) Ganadería ovina sustentable en la Patagonia Austral, tecnología de manejo extensivo. Instituto Nacional de Tecnología Agropecuaria, Argentina, p 270

Briske DD, Derner JD, Brown JR, Fuhlendorf SD, Teague WR, Havstad KM (2008) Rotational grazing on rangelands: reconciliation of perception and experimental evidence. Rangeland Ecol Manag 61:3–17

Butorovic N (2019) Comportamiento de las variables precipitación y temperatura del aire en la ciudad de Punta Arenas durante el período Enero-Julio 2019. Informe solicitado por Empresa Pecket-Energy, Santiago

Castro A (2023) El crecimiento aéreo radical de *Festuca gracillima* Hook. f. (coirón fueguino) y *Holcus lanatus* L (heno blanco). en distintos estados de desarrollo en condiciones controladas. Tesis de pregrado inédita Universidad de Magallanes, Punta Arenas

Collantes MB, Faggi AM (1999) Los humedales del sur de Sudamérica. In: Malvárez AI (ed) Tópicos sobre humedales subtropicales y templados de Sudamérica. UNESCO, Montevideo, pp 15–25

Coronato F (1999) Environmental impacts on offspring survival during the lambing period in Central Patagonia. Int J Biometeorol 43:113–118. https://doi.org/10.1007/s004840050125

Covacevich N (2001) Guía de manejo de coironales: Bases para el planeamiento de la estancia. Boletín INIA 47:1–23

Covacevich N (2012) Sistemas de pastoreo en Magallanes. In: Strauch Ó, Lira R (eds) Bases para la producción ovina en Magallanes. Boletín INIA N° 244, p 152

De Queiroz R, Domingues C, Canozzi M, Garcia R, Ruviaro C, Barcellos J, Rossi J (2018) How do Brazilian citizens perceive animal welfare conditions in poultry, beef, and dairy supply chains? PLoS One 13(12):e0202062. https://doi.org/10.1371/journal.pone.0202062

Dittmann MT, Hummel J, Runge U, Galeffi C, Kreuzer M, Clauss M (2014) Characterising an artiodactyl family inhabiting arid habitats by its metabolism: low metabolism and maintenance requirements in camelids. J Arid Environ 107:41–48. https://doi.org/10.1016/j.jaridenv.2014.04.005

Dwyer C, Lawrence A (2008) Chapter 1: introduction to animal welfare and the sheep in Dwyer C.M. In: The welfare of sheep, vol 6. Editorial Springer Science + Business Media B.V, pp 14–17

Fernández T, Lancaster A, Moraga CA, Radic-Schilling S, von Hardenberg A, Corti P (2021) Behavioural indicators of intra- and inter-specific competition: Sheep co-grazing with guanaco in the Patagonian steppe. Animals, 11: 3333. https://doi.org/10.3390/ani11113333

Egwu G (1992) Outbreak of ovine infectious kerato-conjunctivitis caused by mycoplasma conjuctivae. Small Rumin Res 9:189–196

Endlicher W, Santana A (1988) El clima del sur de la Patagonia y sus aspectos ecológicos. Un siglo de mediciones climatológicas en Punta Arenas. In Anales del Instituto de la Patagonia

Filipová L, Hédl R, Covacevich N (2010) Variability of soil types in wetland meadows in the south of the Chilean Patagonia. Chile J Agric Res 70:266–277

Filipová L, Hédl R, Covacevich N (2013) Magellanic wetlands: more than moor. Folia Geobot 48:163–188

Fulkerson WJ, Donaghy DJ (2001) Plant-soluble carbohydrate reserves and senesces—key criteria for developing an effective grazing management system for ryegrass-based pastures: a review. Aust J Exp Agric 41:261–275

Gammon DM (1978) A review of experiments comparing systems of grazing managment on natural pastures. Proceedings of the Annual Congresses of the Grassland Society of Southern Africa 13:75–82

González BA, Acebes P, Corti P, Grimberg M, Iranzo E, Malo JE et al (2022) Historical perspective and current advances in ecology, conservation, and management of the guanaco in the Chilean Patagonia. In: Carmanchahi PD, Lichtenstein G (eds) Guanacos and people in Patagonia: a socioecological approach to a relationship of conflicts and opportunities. Natural and Social Sciences of Patagonia book series. Springer Nature, Cham

García-Favre J, López IF, Cranston LM, Donaghy DJ, Kemp PD, Ordóñez IP (2023) Functional contribution of two perennial grasses to enhance pasture production and drought resistance under a leaf regrowth stage defoliation criterion. J Agron Crop Sci 209:144–160. https://doi.org/10.1111/jac.12602

Hacker RB (2010) Extensive grazing systems. In: Cottle DJ (ed) International sheep and wool handbook. Nottingham University Press, Nottingham, pp 507–532

Hepp C, Stolpe N (2014) Caracterización y propiedades de los suelos de la Patagonia occidental (Aysén). Instituto de Investigaciones Agropecuarias, Centro de Investigación INIA, Tamel Aike, Coyhaique, Aysén-Patagonia, Chile, p 160

INE (2017) Encuesta de ganado ovino 2017. Instituto Nacional de Estadística Chile [Online] Disponible en. https://bcn.cl/3cdde

Ivelic-Sáez J, Dörner J, Arumí JL, Cisternas L, Valenzuela J, Muñoz E, Clasing R, Valle S, Radic S, Alonso H, López R, Uribe H, Muñoz R, Ordoñez I, Carrasco J (2021) Balance hídrico de humedales de uso agropecuario: El primer paso para el mejoramiento en la gestión hídrica a nivel predial en Magallanes. In: Una investigación multidisciplinaria. Centro Regional de Investigación Kampenaike. Boletín INIA N°435, 162 p, Punta Arenas

Jara R (2018) Factores de demanda de macronutrientes minerales, en las praderas de vega, Tierra del Fuego. Magallanes. Tesis de Magister. Universidad Austral de Chile, Valdivia

Lara A y Cruz G (1987) Vegetación del área de uso agropecuario de la XII Región, Magallanes y la Antártica Chilena, Instituto de Investigaciones Agropecuarias. Santiago

Lancaster A, Corti P, Fernández T, Moraga CA, Radic-Schilling S, von Hardenberg A (2022) Changes in foraging behaviour suggest competition between wild and domestic ungulates: guanaco and domestic sheep in southern Patagonia. J Zool 317(3):213–228. https://doi.org/10.1111/jzo.12971

Lira R (2012) Caracterización del sistema de producción ovina Austral. In: Strauch O, Lira R (eds) Bases para la producción ovina en Magallanes. Centro Regional de Investigaciones Kampenaike, Punta Arenas, Chile, pp 39–47

Luzio W, Casanova M (2006) Avances en el conocimiento de los suelos de Chile. Departamento de Ingeniería y Suelos. Facultad de Ciencias Agronómica. Universidad de Chile; Servicio Agrícola Ganadero (SAG), p 393

Manteca X (2012) Bienestar animal. ANAPORC. Departamento de Ciencia Animal y de los Alimentos. Facultad de Veterinaria, vol 188. Universitat Autónoma de Barcelona, pp 87–97

Manteca X, Temple D, Mainau E, Llonch P (2017) Evaluación del dolor en el ganado ovino. N°17

Marcos MS, Carrera AL, Bertiller MB, Olivera NL (2020) Grazing enhanced spatial heterogeneity of soil dehydrogenase activity in arid shrublands of Patagonia, Argentina. J Soils Sediments 20:883–888. https://doi.org/10.1007/s11368-019-02440-1

Martinic M (2006) El poblamiento rural en magallanes durante el siglo xx.: realidad y utopía. Magallania (Punta Arenas) 34(1):5–20

Navarro M (2021) Determinación de la tasa de crecimiento de especies en diferentes comunidades vegetales presentes en la región de Magallanes y la Antártica Chilena. Trabajo de titulación para optar al título de Ingeniera Agrónoma. Universidad de Magallanes, Facultad de Ciencias Agropecuarias y Acuícolas, Punta Arenas, p 78

Ohrens O, Tortato FR, Hoogesteijn R, Sarno RJ, Quigley H, Goic D, Elbroch LM (2021) Predator tourism improves tolerance for pumas, but may increase future conflict among ranchers in Chile. Biol Conserv 258:109150. https://doi.org/10.1016/j.biocon.2021.109150

Oliva G, Noy Meir I, Cibils A (2001) Fundamentos de ecología de pastizales. In: Borrelli P, Oliva G (eds) Ganadería Ovina Sustentable en la Patagonia Austral. Centro Regional Patagonia Sur, p 272

Ordoñez IP, Oyaneder P, Radic-Schilling S, Valenzuela J, Ivelic-Sáez J, Yagello J (2022) Respuesta fisiológica de Festuca gracillima a la altura de defoliación y estrés hídrico severo. Informativo INIA Kampenaike. N° 119, Punta Arenas

Ordóñez I, Radic-Schilling S, Ivelic-Sáez J, Muñoz-Arriagada R, Pinochet D, Covacevich N, Valle S, Tapia A, Oyaneder P, Valenzuela J, Castro A, Navarro M (2023) Descripción de la vegetación, suelo y tasas de crecimiento de vegas y coironales en la Región de Magallanes. Instituto de Investigaciones Agropecuarias. Boletín N°486, p 100

Ormaechea S, Peri P, Anchorena J, Cipriotti P (2014) Pastoreo estratégico de ambientes para mejorar la producción ovina en campos del ecotono bosque-estepa en Patagonia Sur. Revista Argentina de Producción Animal 34:9–21

Otero L (2006) La huella del fuego. Historia de los bosques nativos. Poblamiento y cambios en el paisaje del sur de Chile. Pehuen, Santiago, p 171

Oyaneder P (2023) Respuesta a la intensidad de defoliación y estrés hídrico de Festuca gracillima Hook, f. bajo condiciones de invernadero. Tesis de pregrado inédita Universidad de Magallanes, Punta Arenas, p 57

Pisano E (1973) Fitogeografía de la Península de Brunswick, Magallanes. Comunidades meso-higromórficas e higromórficas. Anales Instituto Patagonia (Chile) 4:141–206

Pisano E (1977) Fitogeografía de Fuego-Patagonia chilena. I.-Comunidades vegetales entre las latitudes 52 y 56° S. In Anales del Instituto de la Patagonia

Pontigo F, Radic S, Moraga CA, Pulido RG, Corti P (2020) Midsummer trophic overlap between guanaco and sheep in Patagonian rangelands. Rangel Ecol Manag 73(3):394–402. https://doi.org/10.1016/j.rama.2020.01.006

Radic S, Opazo S, Mihovilovic E and Olave C (2011) Determinación de evapotranspiración potencial con los métodos de Penman-Monteith, Ivanov y Turc. Proceedings Congreso Anual de la

Sociedad Chilena de Producción Animal (SOCHIPA), pp 141–142. XXXVI Reunión anual. Punta Arenas, Chile: INIA-UMAG

Radic S, Fernandez A, Opazo S, McAdam J, Ivelic J (2013) Soil bulk density from grasslands in the Magallanes region, Chile. In: 10th International Conference of Agrophysics. Book of Abstract, Lublin, p 105

Radic-Schilling S, Corti P, Muñoz-Arriagada R, Butorovic N, Sánchez-Jardón L (2021) Ecosistemas de estepa en la Patagonia chilena: distribución, clima, biodiversidad y amenazas para su manejo sostenible. In: Castilla JC, Armesto JJ, Martínez-Harms MJ (eds) Conservación en la Patagonia chilena: evaluación del conocimiento, oportunidades y desafíos. Ediciones Universidad Católica, Santiago, pp 223–255

Radic-Schilling S, Sales F, Lira R, Muñoz-Arriagada R, Corti P, Covacevich N, Ivelic-Sáez J, Ordóñez I, Vidal O, Echeverría R, Sandoval C (2022) Magallanes sheep farming. In: Gonzalez M, Palacios C (eds) Sheep farming—herds husbandry and management system, reproductive and improving animal health. IntechOpen, pp 1–29

Radic-Schilling S, Ivelic-Sáez J, Gross P, Ruiz R Muñoz R (2020) Guía de control *Hieracium pilosella* para la Región de Magallanes y la Antártica Chilena, 44 p

Reich PB (2014) The world-wide "fast-slow" plant economics spectrum: a traits manifesto. J Ecol 102:275–301. https://doi.org/10.1111/1365-2745.12211

Romero O, Bravo S (2012) Alimentación y nutrición en los ovinos. In: Romero YO, Bravo MS (eds) Fundamentos De La Producción Ovina En La Región de La Araucanía. Instituto de Investigaciones Agropecuarias, Temuco, pp 23–40

Rozzi R, Armesto JJ, Gutiérrez J, Massardo F, Likens G, Anderson CB, Poole A, Moses K, Hargrove G, Mansilla A, Kennedy JH, Willson M, Jax K, Jones C, Callicott JB, Kalin MT (2012) Integrating ecology and environmental ethics: earth stewardship in the southern end of the Americas. Bioscience 62(3):226–236

Sáez C (1994) Caracterización de la Fertilidad de los suelos de la Región de Magallanes. Características físico-químicas y nutricionales de los suelos. In: Informe Final: Proyecto Fundación Fondo Investigaciones Agropecuarias, Universidad de Magallanes. Escuela de Ciencias y Tecnologías en Recursos Agrícolas y Acuícolas, Punta Arenas

Sales F, Peralta OA, Narbona E, McCoard S, Lira R, De Los RM, González-Bulnes A, Parraguez VH (2019) Maternal supplementation with antioxidant vitamins in sheep results in increased transfer to the fetus and improvement of fetal antioxidant status and development. Antioxidants (Basel) 8(3):59. https://doi.org/10.3390/antiox8030059. PMID: 30857206; PMCID: PMC6466585

Sandoval C, Sales F (2022) Evaluación de la capacidad diferencial materna de producir crías SGA y No-SGA bajo condiciones naturales de restricción nutricional durante la gestación en ovinos de Magallanes. En: Libro de Resúmenes XLVII Congreso SOCHIPA A.G., 29–30 Noviembre, 1 Diciembre, pp 68–69

SENASA (2015) Manual de Bienestar Animal. Servicio Nacional de Sanidad y Calidad Agroalimentaria (SENASA), Argentina, Octubre 2015

Soto LM (1984) Descripción de las praderas naturales de Magallanes continental. I. Área cubierta y composición botánica. Agricultura Técnica 44:185–193

Soto N (2019) Perros bravíos en Magallanes, contexto y marco normativo. In: Comisión de Agricultura, Senado de la República de Chile; 21 January 2019; Valparaíso, Chile

Strauch O, Lira R (2012) Bases para la producción ovina en Magallanes. Boletín INIA 244, Chile, 152 p

Tedla M, Berhan N, Molla W, Temesgen W, Alemu S (2018) Molecular identification and investigations of contagious ecthyma (Orf Virus) in small ruminants, Nort west Ethiopia. BMC Vet Res 2018:14

Turner LR, Donaghy DJ, Lane PA, Rawnsley RP (2006a) Effect of defoliation management, based on leaf stage, on perennial ryegrass (*Lolium perenne* L.), prairie grass (*Bromus willdenowii* Kunth.) and cocksfoot (*Dactylis glomerata* L.) under dryland conditions. 1. Regrowth, tillering and water-soluble carbohydrate concentration. Grass Forage Sci 61:164–174

Turner LR, Donaghy DJ, Lane PA, Rawnsley RP (2006b) Effect of defoliation management, based on leaf stage, on perennial ryegrass (*Lolium perenne* L.), prairie grass (*Bromus willdenowii* Kunth.) and cocksfoot (*Dactylis glomerata* L.) under dryland conditions. 2. Nutritive value. Grass Forage Sci 61:175–181

Turner LR, Donaghy DJ, Lane PA, Rawnsley RP (2006c) Changes in the physiology and feed quality of praire grass during regrowth. Agron J 98:1326–1332

Turner LR, Donaghy DJ, Lane PA, Rawnsley RP (2006d) Effect of defoliation interval on water-soluble carbohydrate and nitrogen energy reserves, regrowth of leaves and roots, and tiller number of cocksfoot (*Dactylis glomerata* L.) plants. Aust J Agric Res 57:243–249

Valle S, Radic S.y, Casanova M (2015) Suelos asociados a tres comunidades vegetales de pastoreo importantes en Patagonia Sur. Revista Agro Sur. 43(2): 89–99

Uzal FA, Songer JG (2008) Diagnosis of *Clostridium perfringens* intestinal infections in sheep and goats. J Vet Diagn Invest 20(3):253–265

Verified Market Reports (2022) Global frozen lamb market by type (lamb head, lamb rack), by application (hypermarkets and supermarkets, independent retailers), by geographic scope and Forecast Available at https://www.verifiedmarketreports.com/product/frozen-lamb-market/)

Villa M, Opazo S, Moraga CA, Muñoz-Arriagada R, Radic S (2020) Patterns of vegetation and climatic conditions derived from satellite images relevant for sub-antarctic rangeland management. Rangel Ecol Manag 73:552–559

Von Moltke C (2009) Distribución *Hieracium pilosella* L en la Provincia de Magallanes. Tesis Pregrado. Universidad de Magallanes, p 50

West DM, Bruère AN, Ridler AL (2009) The sheep: health, disease & production. The sheep: health, disease & production

Xercavins A (1984) Notas sobre el clima de Magallanes (Chile). Revista de geografía, pp. 95–110

Chapter 9
Sheep Production in Southern Chile Based on Pastures

Francisco Canto, José Alfonso Abecia, María Eugenia Martínez, Rodrigo Morales, and Germán Holmberg

Abstract In Chile, there are about 2.5 million head of sheep. The Los Lagos Region accounts for 9% of the national sheep population. Most sheep production systems in this area of Chile are managed by family-type farmers. The representative local breeds are Creole breeds of Iberian origin (Churra and Castellana), including the Künko and Chilote breed. The soil and climatic conditions of this region support the growth of temperate grasslands. Therefore, sheep feeding is based on grazing naturalized grasslands. The botanical composition of the grasslands can vary according to the level of intervention and management. Some pastures are naturalized, while others have been improved by fertilization or drainage. Existing naturalized grasslands have a significant component of scrub and herbaceous species of low forage value. These grasslands have mainly originated from the elimination of wooded areas through logging to transform these lands into soils for agricultural and livestock use. Forage supplementation is carried out exclusively during the winter using hay or grass silage. Meanwhile, concentrate supplementation is used strategically during flushing, the last month of gestation, and the beginning of lactation. Breeding is done through natural mating with one lambing per

F. Canto (✉)
Instituto Universitario de Investigación en Ciencias Ambientales de Aragón, Universidad de Zaragoza, Zaragoza, Spain

Instituto de Investigaciones Agropecuarias, INIA Remehue, Osorno, Chile
e-mail: francisco.canto@inia.cl

J. A. Abecia
Instituto Universitario de Investigación en Ciencias Ambientales de Aragón, Universidad de Zaragoza, Zaragoza, Spain

M. E. Martínez
Instituto de Investigaciones Agropecuarias, INIA Butalcura, Castro, Chile

R. Morales
Instituto de Investigaciones Agropecuarias, INIA Ururi, Arica, Chile

G. Holmberg
Instituto de Investigaciones Agropecuarias, INIA Remehue, Osorno, Chile

C. Palacios Riocerezo, M. Gonzalez Ronquillo (eds.), *Traditional Livestock Production*, https://doi.org/10.1007/978-3-031-88275-3_9

year. Some farmers schedule lambing to coincide with the period of greatest pasture availability and better environmental conditions. This way, at lambing time, the ewes have access to high nutritional value resources at low cost. Herd health management throughout the year includes hoof trims, antiparasitic treatments, and clostridial vaccine application. It is important to note that Chile is free of major diseases that affect sheep (scrapie, foot-and-mouth disease, and ovine brucellosis). Lambs are slaughtered at 30–35 kg live weight for the year-end holidays. Depending on the conditions, management, and socio-cultural traditions, there are opportunities and potential for the development of organic livestock.

Keywords Production systems · Sheep · Farming · Extensive grazing · Lamb meat · Chile

9.1 Introduction

Sheep farming in Southern Chile is an evidence of centuries of adaptation, resilience, and traditional socio-agricultural practices. Sheep were first introduced to the Los Lagos Region during the Spanish colonial period, and they have prospered in its diverse environments, from the Andean foothills to the valleys and coastal plains of the region.

The history of sheep farming in southern Chile begins with the arrival of Churro-trunk sheep from Peru during the sixteenth century, initiating livestock practices that have influenced local economies and cultural traditions (Carvallo 1875; Miller 1980). Initially introduced via maritime routes to the Valdivia and Chiloé Island ports (Gay 1862; De La Barra et al. 2018), sheep became integral to the livelihoods of Mapuche-Huilliche communities, adapting to the distinctive climatic and ecological conditions of the region.

Over time, sheep production in Southern Chile has evolved under extensive management systems, utilizing marginal lands with limited pasture resources and facing the challenges posed by the variable weather patterns of the region (Lobos and Pávez 2021). The resilience of local sheep breeds, shaped by centuries of natural selection and genetic adaptation, has been crucial in sustaining production amidst these conditions (De La Barra et al. 2011).

Sheep farming in Southern Chile remains characterized by its support on traditional practices handed down through generations, while also incorporating modern techniques and innovations to enhance productivity and sustainability. This chapter provides a characterization of sheep farming systems, including husbandry practices, under the soil and climatic conditions in Los Lagos Region, south of Chile.

9.2 Geography and Climate

The Los Lagos Region covers from the Argentine Republic border to the Pacific Ocean, between 40°15′ to 44°14′. These reliefs show a multitude of geological processes and forms. Most of the actual formations are the result of several historical processes, with the action of ice and volcanic activity as protagonists (Butland 1957).

For descriptive purposes in this chapter, the region is divided into three zones according to sheep production systems (Fig. 9.1): Continental North, Chiloé archipelago, and Patagonia Verde Territory. In the first zone, to the north of the city of Puerto Montt, geography is strongly influenced by lakes, including in an east-west direction the Andes mountain range, the Central Valley, and the Coastal Mountains (what is known as the Continental North). Towards the south of Puerto Montt, the Central Valley comes to an abrupt end due to tectonic subsidence towards the south of the Gulf of Reloncaví and the Chacao Channel. The second zone corresponds to the set of islands that make up the Chiloé archipelago, while the third zone includes the so-called Patagonia Verde territory.

Most of the soils in the region are of volcanic origin. In the Continental North there are areas defined according to their agricultural suitability: while the soils of the Andes Mountains are suitable for forestry, the best agricultural lands are found in the center of the Central Valley, with cereal crops and cattle livestock (meat and dairy). Towards the coast, there are red clay soils that are characterized by moderate to low fertility (Diaz et al. 1960). Therefore, most of the sheep production systems in this area are mainly located in the coastal mountain range and the Andean foothills (Desauguste et al. 2011) (Fig. 9.2).

The different genesis of the soils of the Chiloé archipelago has resulted in low pH levels, high levels of aluminum saturation, high contents of organic matter and low levels of bases and elements. This has defined a slow development of the vegetation cover and a low nutritional value for livestock (De La Barra et al. 2011).

Patagonia Verde is characterized by low fertility soils, pastures with low forage value and short growing period, high-slope grazing areas and a harsh geography (Lobos and Pavéz 2021).

The Los Lagos region is characterized by a humid temperate climate, classified as Marine West Coast (Sarricolea et al. 2017). Precipitation fluctuates between 1600 and 2500 mm/year (depending on the area), being one of the highest in the country. The average temperatures in the winter months are no lower than 6 °C, and snow is a rare phenomenon in both the Central Valley and the Coast, thus allowing for some winter growth of the grasslands. The environmental humidity is relatively high during most of the months, mainly between March and October, due to the fact that during this period the greatest amount of water falls during the year.

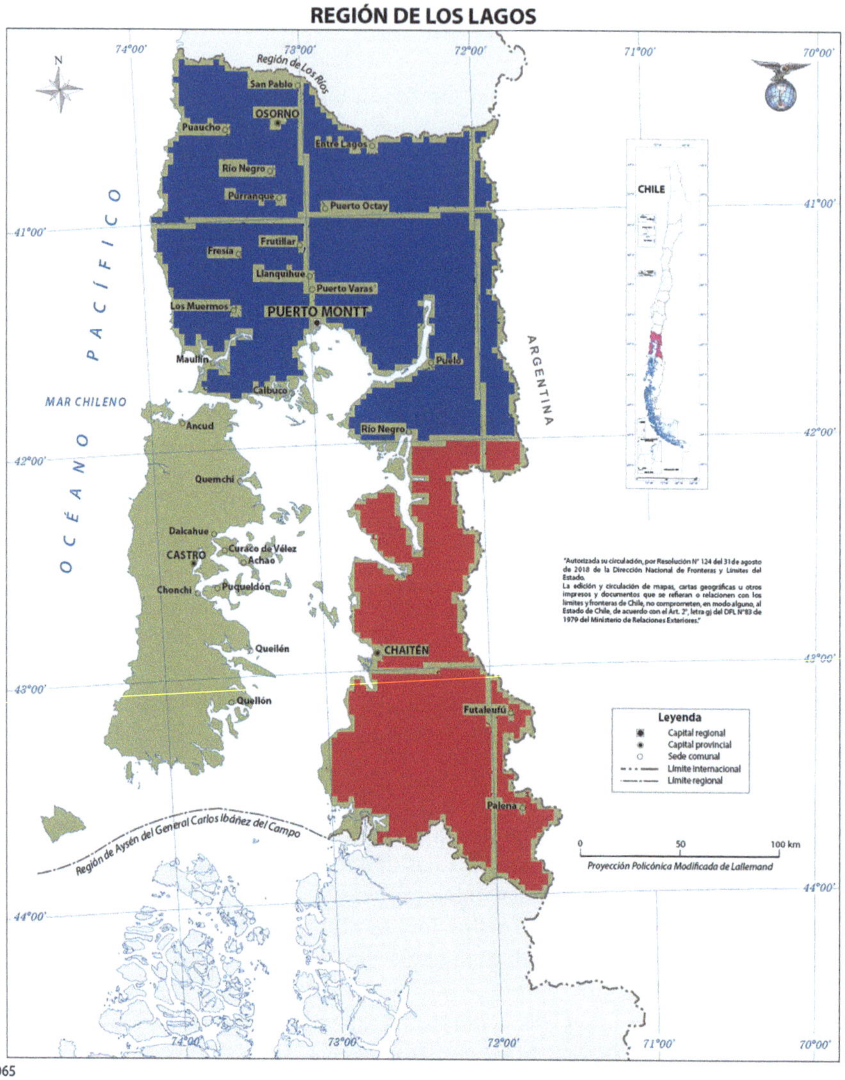

Fig. 9.1 Schematic distribution of the three sections in Los Lagos (blue: Continental north; green: Chiloé archipielago; red: Patagonia Verde). Adapted from Geographic Military Institute of Chile

9.3 Sheep Census

In Chile, approximately half of lamb meat production is exported. The rest corresponds to national consumption, both formal and informal, so it is difficult to quantify (ODEPA 2014). According to the latest national agricultural census (INE 2021), there are 2,529,238 sheep in Chile. Compared to the previous census (INE 2007),

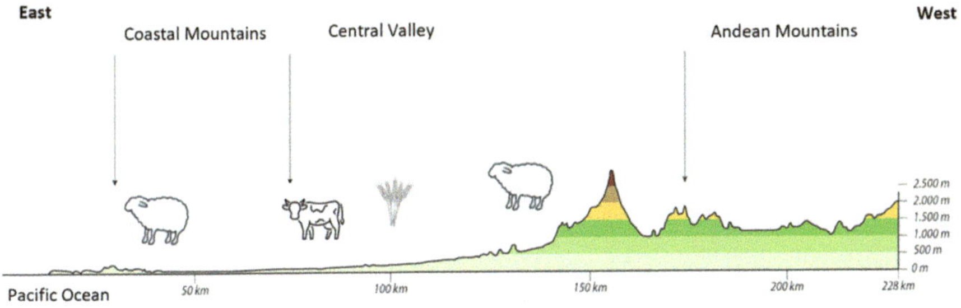

Fig. 9.2 Illustrative topographic profile of the Continental North area in Los Lagos Region, Chile. Adapted from Geographic Military Institute of Chile

the national sheep population decreased by approximately 35%. The sheep sector has a high geographic concentration in the Magallanes Region (Southern Patagonia), currently reaching 56% of the total population. Los Lagos region follows, with 9% (227,798 sheep). 24.7% of Chilean sheep producers are in this region, were sheep farming plays an important role in the development of the family economy, providing an important source of incomes and requiring a lower amount of labor and investment compared to other livestock (Moscoso et al. 2021).

9.4 Sheep Farmers

The low average number of sheep in most regional flocks (less than 50 heads and less than 50 hectares) allows them to be classified as family-type farms, oriented toward self-consumption, with occasional sales.

In the Los Lagos region, the sociodemographic composition reflects national trends. Most of the breeders are men (78.7%) (INE 2017). The number of farmers dedicated to this activity has decreased by 20.5% compared to the last national census (2007). The group is characterized by the relatively old age of the farmers (58% are over 61 years old). This is increasingly common, given the decline in sheep farms because of the low profitability of sheep production and the migration of the young population to the cities (Larrondo et al. 2018; Toro-Mujica and Riveros 2021).

Smallholder farmers often lack basic knowledge of sheep management. The cost of supplementary food and medicines, the lack of veterinary care in the most remote areas of the region, and the malnutrition are the main causes of perinatal deaths. In general, there are low birth and weaning weights of lambs, which result in low productivity and significant financial losses (De La Barra and Bravo 2006). Despite encountering production limitations, farmers in this region have successfully developed models of associativity and cooperativism aimed at sustainable agricultural and livestock development. As a result, the Sociedad de Fomento Ovejero de Chiloé (established in 2012) and Cooperativa Ovinos San Juan (established in 2019) have been created in the Chiloé Archipelago and Continental North areas, respectively. They work collaboratively with farmers and public-private institutions, thereby facilitating the development of innovation projects through teamwork.

9.5 Sheep Breeds

Several sheep breeds have been introduced into Chile since the 1960s. In Los Lagos region, the most widely distributed sheep breed is the Romney Marsh, a rustic and dual-purpose animal, well fitted to humid regions and less susceptible to parasitic diseases, typical of the environmental conditions found in Southern Chile (Mujica 2005). The second breed in number is the Suffolk Down, being preferably used for crossbreeding with other breeds to improve the meat quality of the flocks (Leymaster 2002). There are also local breeds that have been generated from an adaptive process and the selection schemes applied by the rural communities associated with the territories (De La Barra et al. 2016).

The Chilota sheep breed is found on the Chiloé archipelago (De La Barra et al. 2010). It is a descendant of the Iberian churro line, and it is adapted to local environmental conditions (Farias et al. 2010; De La Barra et al. 2011). The Chilota breed has low body weight, suggesting processes of adaptation to low temperature conditions and shortages of food. Also, Chilota's rusticity via resistance to Foot Rot and gastrointestinal parasites has been reported (Martínez et al. 2012).

The Künko sheep breed, also of Iberian origin, is found in the coastal area between the Maullín River and Pilmaiquén, but with the incorporation of Hampshire-down genetics to some extent in the last 100 years (De La Barra et al. 2016) (Fig. 9.3). This breed is characterized by being dual-purpose (meat and wool), of great rusticity and maternal ability, and raised mainly by indigenous people of the Mapuche Huilliche ethnic group.

In Patagonia Verde (communes of Hualaihue, Cochamó, Chaitén, Palena and Futaleufú) the largest proportion of the flocks correspond to hybrid animals that are not assignable to any particular racial biotype. Hybridization is higher in the most isolated communes and lower in the most accessible ones (De la Barra et al. 2019).

9.6 Husbandry Practices

The practices employed by sheep farmers differ considerably across regions. For this reason, only the main technological practices that improve productivity will be discussed in this chapter (Peña et al. 2014).

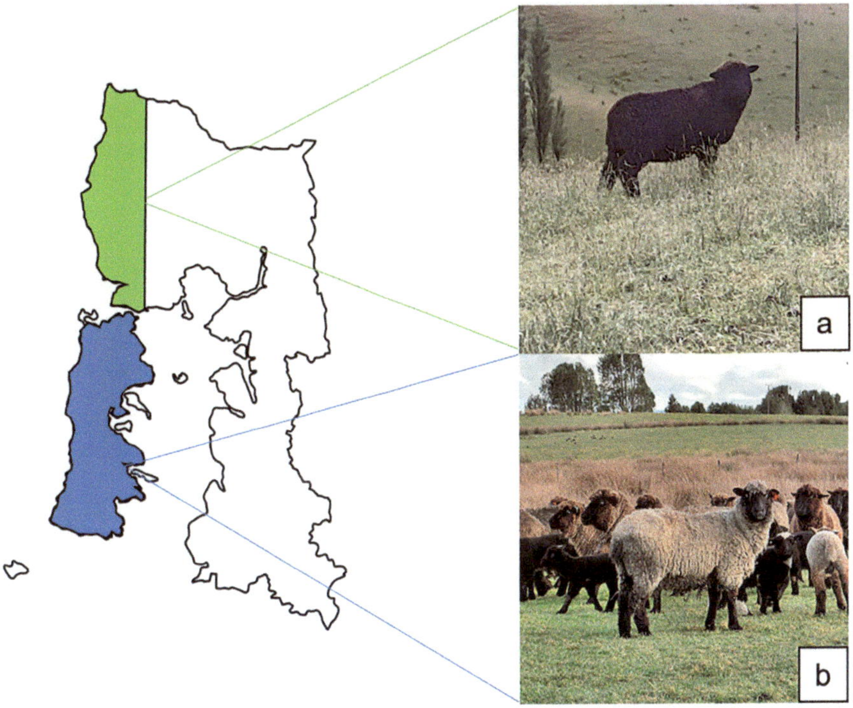

Fig. 9.3 Geographic distribution of the main Creole sheep breeds in Los Lagos region, Chile (a = Künko; b = Chilota) (Photo credits: Künko sheep by Félix Cancino and Chilota sheep by Luis Opazo))

9.6.1 Health Status and Sanitary Management

Chile is free of the main livestock diseases such as foot–mouth disease (since 1981), scrapie (since 2013) and caprine and ovine brucellosis (*Brucella melitensis*) (since 2013) (SAG 2021). In addition, the country employs an Exotic Disease Surveillance Plan, which includes bluetongue monitoring of susceptible animals in risk areas to support the absence of viral circulation (SAG 2021). No cases of Maedi Visna have been reported in Patagonia Verde (SAG 2020). This sanitary condition and the control mechanisms allow achieving higher production levels in the flocks, lower costs associated with vaccines and medications, and market potential in international markets (Gallo et al. 2018).

As mentioned above, sheep feed mainly on grasslands. However, in some periods of the year the diet is complemented with preserved forage and/or concentrated feed. This can increase the risk of clostridial infections (Clostridium perfringens types C and D) in animals (Fig. 9.4). For this reason, farmers are recommended to vaccinate sheep against clostridial diseases 30 days before lambing and lambs in the third and seventh week after birth (Calderón and Martínez 2015).

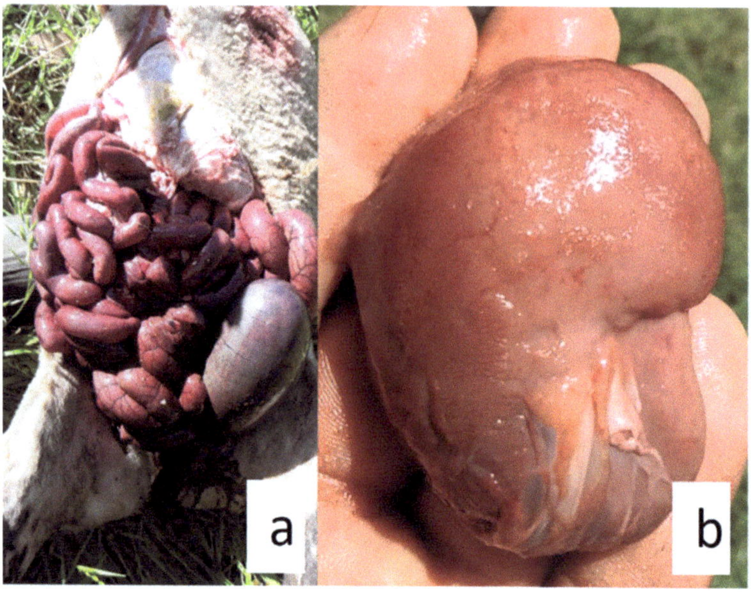

Fig. 9.4 Post-mortem lesions of the main clostridial diseases in southern Chile. (**a**) Small intestine of a lamb with hemorrhagic enteritis due to Clostridium perfringens type C; (**b**) Pulpy kidney due to Clostridium perfringens type D

The high rainfall and the night confinement to which most flocks are subjected contribute to the formation of mud and the accumulation of urine and feces in the soil. These factors create an unfavorable environment for the health of the hooves (Alomar 1998). The most common foot pathologies are oversized hooves, hoof separation and diffuse white line lesions (Tadich et al. 2000). Although foot diseases are highly prevalent in flocks, hoof trimming is used as a treatment for lameness and not preventively as part of a flock health program (Tadich et al. 2000; Canto 2017). Preventive hoof trimming should be carried out twice a year (prior to mating and at shearing) throughout the entire flock. To perform it, different types of cutters and copper-based disinfectant solutions are used (Canto et al. 2019) (Fig. 9.5).

9.6.2 Pasture and Feeding Management

The sustainability of a livestock system depends on many factors. Probably the most important factor is ensuring that animals have enough feed to meet their maintenance and production requirements. In the sheep systems of Southern Chile, sheep feeding is based on the use of pastures, providing a low-cost supply of nutrients throughout the year (Canto 2017). This forage resource can contribute up to 90% of the sheep annual ration (Fig. 9.6). The plant community of naturalized grasslands

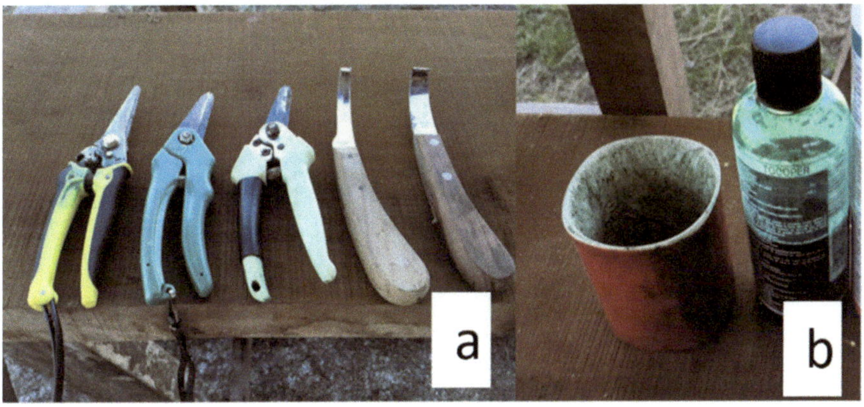

Fig. 9.5 Tools (**a**) and copper solution (**b**) used for this hoof trimming in Southern Chile

Fig. 9.6 Contribution of different feedstuffs for sheep in Chiloé. (Credits of picture Paulo Guarda)

varies depending on the level of intervention and management. Some of the common species are *Holcus lanatus, Bromus Valdivian, Bromus stamineus, Lotus uliginosus* and *Plantago lanceolata* (Ramirez et al. 1991).

The two main types of pastures are the natural grasslands (5957 kg Dry Matter (DM)/ha/year) and those improved through fertilization or drainage (9934 kg DM/ha/year); they are mainly composed of grasses and legumes (Moscoso et al. 2021) (Fig. 9.7). The productivity and nutritional quality of grasslands relies on local climatic conditions, water availability, soil fertility, forage species, use and management.

Sheep systems in the region must be able to establish a balance between pasture production (kg DM/ha) and the number of animals per hectare (stocking rate). To achieve an adequate adjustment, the consumption capacity of the sheep (which depends among other factors, on live weight) and the utilization efficiency, which depends on the grazing system used, must be considered. Table 9.1 summarizes the different stocking rates supported by a pasture of 6000 kg DM/ha, depending on the breed and the different grazing management (Canto et al. 2021).

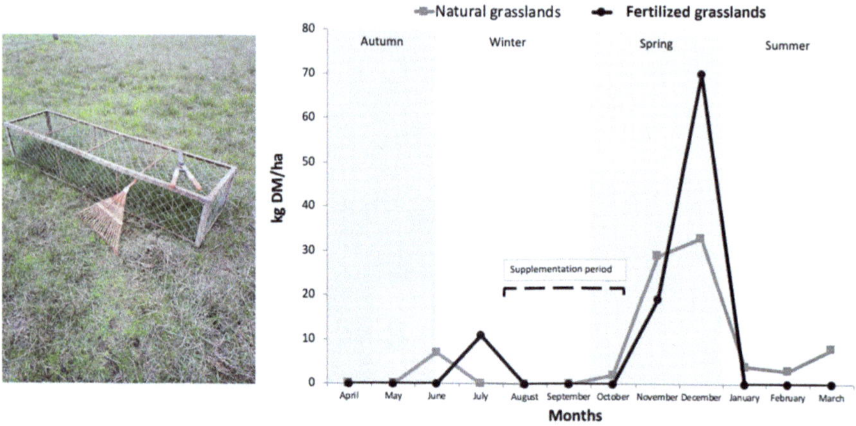

Fig. 9.7 Exclusion cage method (left) used to estimate the growing rate (kg DM/day) of natural and fertilized grasslands (right) in Chilóe archipelago, Chile (Moscoso et al. 2021)

Table 9.1 Adjustment of stocking rate (sheep/ha) according to breed and grazing method, based on pasture production of 6.000 kg DM/ha/year in Los Lagos Region, Chile

	Stocking rate capacity (sheep/ha)		
Grazing type	Chilota or Künko	Romney Marsh or creole in breeding	Suffolk down
Continuous grazing	5.0	4.0	3.0
Rotative grazing	6.0	5.0	4.0
Strip grazing with electric fence	7.0	6.0	5.0

Although each type of grass has a particular production curve, most of them show a clear seasonality, characterized by high production rates during spring, which decrease in summer and winter. The harshness of the summer period is related to water deficit, and the winter period to the incidence of low temperatures and less solar radiation (González and Tapia 2017). In the annual mating scheme usually used, sheep get pregnant in the autumn, the pastures do not provide sheep with the necessary quantity and quality of nutrients during the last third of gestation (between June and August in the Southern hemisphere), and some farmers give an strategic supplementation to the sheep (Canto and Martínez 2018) (Fig. 9.7). Supplementary feeds usually include between 20 and 25 kg/sheep/year of grain (mainly oats) and/or pelleted concentrates. The administration of these food supplements to sheep at the time of greatest requirements has a positive effect on the birth weight of the lamb and the subsequent growth rate (Martínez et al. 2012, 2015).

Fig. 9.8 Ram coloring scheme for mating programming and estimation of probable lambing date

9.6.3 Reproductive Management

Sheep are seasonally polyestrous. The breeding season is a succession of estrous cycles, with an average length of 17 days. The ovulatory period usually begins in late summer or early autumn in response to shortening day length, and ends in the late winter or early spring (Abecia et al. 2011).

In an extensive sheep production system, mating scheduling is essential, because it allows births to be grouped and to occur when environmental conditions are more favorable, providing a low-cost food resource with high nutritional value. This practice is effective when the ram remains olfactorily and visually separated from the ewes throughout the year. The ram only joins the flock during the mating season, for three whole cycles (~51 days). In Los Lagos region, only 14.8% of sheep farmers apply this practice (INE, 2017). The limited number of paddocks available for rams could be the main cause of this low technology adoption. Some farmers paint the ram's chest with colored chalk and vegetable oil. The color changes every 17 days; this way, farmers can estimate lambing dates and identify the sheep that were not mated (Calderón and De la Barra 2010) (Fig. 9.8).

The majority of sheep farmers (99.4%) carry out natural mating systems with one lambing per year. Breeding occurs between February and April in the continental north, between March and April in Chiloé and between April and May in Patagonia Verde, without hormonal synchronization. Techniques that improve the productivity and genetics of flocks, such as ultrasound or artificial insemination are scarcely used (INE, 2017). However, these extensive and natural management conditions show fertility rates (lambed ewes/mated ewes) above 90% and 92% in the Patagonia Verde and central valley of the Continental North area, respectively (Cox et al. 2015; Canto 2017).

9.6.4 *Manure Management*

Sheep manure is used to elaborate compost. This practice occurs more frequently in the smaller sheep farmers (69% of producers who have a flock of up to 10 ewes) (ODEPA, 2013).

9.7 Products

9.7.1 *Wool*

Sheep wool production in the Los Lagos region amounted 50,320 kg during the agricultural productive year of 2009/2010, representing 8.2% of the national production (INE, 2010). In this region, wool sold via retail markets, with raw white wool prices varying between 300 and 400 Chilean pesos ($ CLP) per kilo. Washed and carded white wool retails for around 10,000 pesos per kilo (Petit-Breuilh et al. 2013).

For the family-type farmers in the southern zone, it is possible to increase profits from sheep farming by utilizing not only meat but also by products such as wool. Accessible and low-investment techniques can be adopted for their implementation. This offers an opportunity to differentiate through the use of colored wools found in Creole breeds (Guarda and Díaz 2016). In response to this opportunity, productive capacities and infrastructure have been developed for the process of conditioning and craft transformation of pigmented sheep wool in the Los Lagos Region (Lobos and Pavéz 2018). As a complement to these advancements, the "Wool Bank in the Los Lagos Region" was established to further support local producers. This initiative aimed to narrow the gap between supply and demand by ensuring access to high-quality wool at fair prices and facilitating equitable market entry for producers (FIA, 2016).

9.7.2 *Meat*

In recent years, there has been a growing interest in meat produced from free-range grazing systems, where animals have access to natural pastures and are not confined to feedlots. Chilean animals raised in free-range grazing systems have been shown to contain lower levels of intramuscular fat and high levels of healthy fats, such as conjugated linoleic acid and a favorable n-6/n-3 ratio. This latter may contribute to reducing the risk of human coronary diseases, making it an appealing option for health-conscious consumers looking for sustainable and ethical meat choices (Morales et al. 2012; Morales and Subiabre 2022).

The Los Lagos region is characterized by being oriented towards meat production with a clear destination for the domestic market, preferably the consumption of the regional population and the consumption associated with tourism (Lobos et al. 2021). Lamb meat consumption in Chile is traditional and remains strongly seasonal, primarily aligning with national holidays in September and December and during the summer vacation period. Lamb meat's seasonal availability limits its consumption throughout the year, especially in a country where certain traditional festivities are closely tied to lamb meat (Sepúlveda et al. 2017).

Both the Geographical Indication (GI) and the appellation of origin (D.O.) in Chile are quality brands that acknowledge the uniqueness of a product when its quality, reputation or other characteristics are fundamentally attributable to its geographical origin. This is the case of Cordero Chilote (GI), which production is historically and closely linked to Chiloé. Meat quality of this lamb meat is derived from (1) the breed; (2) the agroecological conditions in the Chiloé archipelago; and (3) the traditional production practices (suckling and grazing) (Martínez et al. 2017). Other quality brands available in Chile are collective brands that are available to groups of farmers to differentiate their production. In the Los Lagos region, there is the collective brand "ÓvejaK¨" that protects Künko lamb products (wood, meat, milk, etc). The Künko lamb is associated with Huilliche farmers of the coast of the region.

Studies were performed on the nutritional quality of lamb carcasses (Ramirez-Retamal et al. 2013) and meat (Ramirez-Retamal et al. 2014) of two breeds i.e., Chilote and Suffolk down, and fed on naturalized grassland or rangeland pastures. In the carcass measurements, Chilote lamb is narrower than Suffolk Down, but with a higher proportion of hide and hooves. No effects of breed or pasture type on the main characteristics of the lamb carcasses were observed (Ramirez-Retamal et al. 2013). However, the pasture type did affect the meat quality, given that Chilote lambs fed on naturalized pasture had a lower shear force and higher intramuscular fat levels, whereas Chilote lambs fed on rangeland pasture showed higher percentages of n-3, n-6 fatty acids and polyunsaturated fatty acids. The concentrations of fatty acids were similar in both groups. The results indicated some evidence that the type of pasture of Chiloe archipelago affects meat quality that could form the basis to generate a premium product (Ramirez-Retamal et al. 2014).

Table 9.2 Nutritional label of two lambs produced in the Los Lagos region

Parameter	Chilote[a]	Künko[b]
Energy (kcal)	125	140
Protein (g)	18	19
Total fat (g)	5.7	7.0
Saturated fatty acids (g)	2.5	2.2
Monounsaturated fatty acids (g)	2.0	1.7
Polyunsaturated fatty acids (g)	0.5	0.4
Trans fatty acids (g)	0.4	0.3
Cholesterol (mg)	51	60
Total carbohydrate (g)	0.6	1.0
Total sugars (g)	0.4	0.3
Sodium (mg)	52	63

[a]Morales et al. (2017)
[b]Unpublished data

Regarding the Künko lamb meat quality is quite similar to Chilote Lamb as presented in Table 9.2.

9.7.3 *Cheese*

Traditionally, sheep producers in Southern Chile do not milk sheep, since the production system is based on the production of lamb meat. The milk from the sheep is almost exclusively used to breastfeed the lambs. Only in a few cases, sheep farming in the region is intended for production of sheep milk or cheese. One of the brands of sheep cheese found in Lagos region is the Chilozábal cheese made in Chiloé, a mature cheese. This sheep milk cheese simulates the process involved in the making of Idiázabal cheese, a Spanish cheese with indication of origin. Chilozábal cheese is produced using milk from 140 Latxa animals, yielding 700 kg/year (Barrón Rivas et al. 2019).

9.8 Organic Production

In Chile, organic production is regulated by Law 20,089, which is part of the National Organic Agriculture System (Congreso Nacional de Chile, 2015). In Title 6 (articles 20–28), this law establishes the specific rules for organic livestock production. This regulation does not differentiate between the concepts organic, ecological, or agroecological, so for the purposes of this section, they will be used interchangeably.

Organic products are understood to be coming from holistic production management systems in the agricultural, livestock or forestry field, which promote and improve the health of the agroecosystem and biodiversity, biological cycles and the biological activity of the soil. In Chile, meat, eggs, skin, fibers, wool, hair, and primary and processed by-products obtained in accordance with the normative can be certified as organic products (FAO, 2005).

The Agricultural and Livestock Service (SAG) of Chile ensures the compliance with this normative through certification entities called Official Certification Organizations. There exist seven certification entities (SAG, 2024), but only one of them can certify livestock products, given that organic livestock production is still a minority in the country.

9.9 Conclusions

Sheep production in the Los Lagos region is based in raising the animals on pasture-fed, being a seasonal production, with lamb meat as the main product. It is a traditional production-related family-type farm, mainly oriented toward self-consumption. The sheep farming system, as have been described in this chapter, is to a certain extent compatible with organic production. Farmers can make the transition to organic production in a relatively easy way since they are not so far from traditional practices.

As a tourist region that demands differentiated and local origin products, encouraging farmers to obtain organic certification is a feasible strategy to add value to their production. However, before intending a new certification, it would be better to take advantage of the quality brands available in this region, such as Cordero Chilote GI.

References

Abecia JA, Forcada F, González-Bulnes A (2011) Pharmaceutical control of reproduction in sheep and goats. Vet Clin N Am-Food A 27:67–79. https://doi.org/10.1016/j.cvfa.2010.10.001

Alomar D (1998) Producción ovina. In: Universidad Austral de Chile (ed) Pequeña agricultura en la X Región de los Lagos, Chile. Valdivia, pp 119–135

Barrón Rivas MC, Palacios Riocerezo C, Domínguez Vara IA, González Ronquillo M, Radic Schilling S (2019) Production, processing, commercialization and analysis of costumer preferences of sheep cheese in Chile. In: Javed K (ed) Milk production. Processing and Marketing, IntechOpen. https://doi.org/10.5772/intechopen.83806

Butland GJ (1957) The human geography of southern Chile. Institute of British Geographers, N° 24. United Kingdom

Calderón C, De la Barra R (2010) Programación del encaste en rebaños ovinos de pequeñas explotaciones. Osorno, Chile. Instituto de Investigaciones Agropecuarias. Informativo INIA Remehue N° 77, 1-2. Available at https://hdl.handle.net/20.500.14001/4730. Accessed on 16 Feb 2024

Calderón C, Martínez ME (2015) Prevención de enterotoxemia en el ganado ovino. Osorno, Chile. Instituto de Investigaciones Agropecuarias. Ficha técnica INIA Remehue N° 08, pp. 1–2. Available at https://hdl.handle.net/20.500.14001/66836. Accessed on 16 Feb 2024

Canto F (2017) Caracterización de sistemas de producción ovina en pequeños Agricultores de la costa y precordillera en la provincia de Palena. In: Iraira S, Canto F, Siebald E, Vera C (eds) Programa de fertilización de praderas. Líneas de intervención para el mejoramiento productivo de la actividad pecuaria de la provincia de Palena. Osorno, Chile. Instituto de Investigaciones Agropecuarias. Boletín INIA N° 345, pp. 33–55. Available at https://hdl.handle.net/20.500.14001/6584. Accessed on 16 Feb 2024

Canto F, Martínez ME (2018) Suplementación nutricional estratégica en ovinos del sur de Chile. Osorno, Chile. Instituto de Investigaciones Agropecuarias. Informativo INIA Remehue N° 188, pp. 1–2. Available at https://hdl.handle.net/20.500.14001/4902. Accessed on 16 Feb 2024

Canto F, Schwerter X, Vidal D (2019) Enfermedades y manejo podal en ovinos. Osorno, Chile. Instituto de Investigaciones Agropecuarias. Ficha técnica INIA Remehue N° 29, pp. 1–2. Available at https://hdl.handle.net/20.500.14001/66916. Accessed on 16 Feb 2024

Canto F, Moscoso C, Guarda P, Chahín G (2021) Ajuste de carga anual de ovinos. Osorno, Chile. Instituto de Investigaciones Agropecuarias. Ficha técnica INIA Remehue N° 122, pp. 1–2. Available at https://hdl.handle.net/20.500.14001/67496. Accessed on 16 Feb 2024

Carvallo V (1875) Descripción histórico-geográfica del reino de Chile. Librería del Mercurio, Santiago

Congreso Nacional de Chile (2015) Ley 20.089. Sistema Nacional de Certificación de Productos Orgánicos Agrícolas. Available at https://www.bcn.cl/leychile/navegar?idNorma=246460. Accessed on 16 Feb 2024

Cox JF, Jeria E, Bocic A, Soto-Saravia R, Dorado J, Saravia F (2015) Characterization of the productive performance of highlander sheep in Southern Chile. I. Female reproductive traits Small Rumin Res 130:183–188. https://doi.org/10.1016/j.smallrumres.2015.06.010

De La Barra R, Bravo R (2006) La mortalidad ovina en el sur de Chile. Osorno, Chile. Instituto de Investigaciones Agropecuarias. Informativo INIA Remehue N° 60, pp. 1–2. Available at https://hdl.handle.net/20.500.14001/4218. Accessed on 16 Feb 2024

De La Barra R, Uribe H, Latorre E, Arranz JJ, San Primitivo F (2010) Genetic structure and diversity of four Chilean sheep breeds. Chil. J. Agric. Res. 70(4):646–651

De La Barra R, Carvajal A, Uribe H, Martínez ME, Gonzalo C, Arranz J, San Primitivo F (2011) El ovino criollo Chilote y su potencial productivo. Anim Genet Resour 48:93–99. https://doi.org/10.1017/S2078633610000937

De La Barra R, Carvajal A, Martínez ME, Guarda P, Calderón C (2016) Differentiation and morphostructural variability of Künko biotype sheep in Los Lagos region, Chile. Journal of Animal Ethnology 2:1–8

De La Barra R, Martínez ME, Carvajal A (2018) Conservación genética y registro de nuevas razas ovinas en Chile. Actas Iberoamericanas de Conservación Animal 12:9–15

De La Barra R, Carvajal A, Martínez ME, Palavecinos P (2019) Diversidad racial de la ganadería ovina en el Territorio Patagonia Verde, Chile. Actas Iberoamericanas de Conservación Animal 13:41–51

Desauguste M, Lerdon FJ, Moreira LV, Alomar CD (2011) Caracterización de la producción ovina en la agricultura familiar de la comuna de Paillaco, Región de Los Ríos, Chile. Agrosur 39:88–94. https://doi.org/10.4206/agrosur.2011.v39n2-03

Diaz C, Astudillo J, Aranda G (1960) Reconocimiento de suelos de las Provincias de Osorno y Llanquihue. Agricultura Técnica 19–20:125–205. Available at https://hdl.handle.net/20.500.14001/37690. Accessed on 16 Feb 2024

FAO (Food and Agriculture Organization of the United Nations) (2005) Codex Alimentarius 2ª Edition. Available at https://www.fao.org/3/a0369s/a0369s.pdf. Accessed on 16 Feb 2024

Farías E, Mujica F, Hervé M (2010) Diagnosis of the distribution of ovine genetic resources from the Coquimbo region to Magallanes y Antartica Chilena region. Agro Sur 38(2):97–108

FIA (Fundación para la Innovación Agraria) (2016) Ficha Iniciativa FIA: Banco de lanas. En busca de un modelo innovador para mejorar la competitividad del sector textil artesanal tradicional. Available at https://bibliotecadigital.fia.cl/server/api/core/bitstreams/9e94000f-a22c-4996-8dec-aab380d41af9/content. Accessed on 16 Feb 2024

Gallo C, Tarumán J, Larrondo C (2018) Main factors affecting animal welfare and meat quality in lambs for slaughter in Chile. Animals 8:165. https://doi.org/10.3390/ani8100165

Gay C (1862) Historia física y política de Chile. Agricultura. Imprenta E. Thunet y Cía. París, Francia, Tomo I

González MV, Tapia MM (2017) Manual de manejo ovino. Boletín INIA - Instituto de Investigaciones Agropecuarias, Vol. no. 368. 158 p

Guarda P, Díaz I (2016) Mejorando la rentabilidad del negocio ovino. Lanas y cueros una opción. Osorno, Chile. Instituto de Investigaciones Agropecuarias. Informativo INIA Remehue N° 168, 1–2. Available at https://biblioteca.inia.cl/handle/20.500.14001/4795. Accessed on 16 Feb 2024

INE (Instituto Nacional de Estadística) (2007) Censo agropecuario. Base de microdatos. CD del censo agropecuario 2007

INE (Instituto Nacional de Estadística) (2010) Encuesta de Ganado Ovino 2010. Available athttp://www.ine.cl/canales/menu/publicaciones/calendario_de_publicaciones/pdf/230611/ovino_10220611.pdf. Accessed on 16 Feb 2024

INE (Instituto Nacional de Estadística) (2017) Encuesta de Ganado Ovino 2017. Available online at https://www.odepa.gob.cl/wp-content/uploads/2017/01/Encuesta-de-ganado-ovino.xlsx. Accessed on 16 Feb 2024

INE (Instituto Nacional de Estadística) (2021) Censo Agropecuario. Available at https://www.ine.gob.cl/estadisticas/economia/agricultura-agroindustria-y-pesca/censos-agropecuarios#:~:text=Cuadros%20Estad%C3%ADsticos-,2021,-2007. Accessed on 16 Feb 2024

Larrondo C, Bustamante H, Gallo C (2018) Sheep farmers' perception of welfare and pain associated with routine husbandry practices in Chile. Animals 8:225. https://doi.org/10.3390/ani8120225

Leymaster KA (2002) Fundamental aspects of crossbreeding of sheep. Use of breed diversity to improve efficiency of meat production. Sheep Goat Res J 17:50–59

Lobos I, Pávez P (2018) Tecnificación del proceso de acondicionamiento y transformación artesanal de Lanas y Cueros Ovinos pigmentados en la región de Los Lagos. Osorno, Chile. Instituto de Investigaciones Agropecuarias. Boletín INIA N° 364. Available at https://hdl.handle.net/20.500.14001/6717. Accessed on 16 Feb 2024

Lobos I, Pávez P (2021) Agregación de valor del ovino en el territorio Patagonia Verde, región de Los Lagos. Osorno, Chile. Instituto de Investigaciones Agropecuarias. Boletín INIA N° 456. Available at https://hdl.handle.net/20.500.14001/68370. Accessed on 16 Feb 2024

Martínez ME, Calderón C, De La Barra R, Uribe H (2012) Effect of management practices in the productive performance of three sheep breeds in the Chiloé archipelago. Chile Livest Sci 3:57–66

Martínez ME, De la Barra R, De la Fuente F (2015) Effect of early creep feeding in Chilota breed lambs. Livest Sci 6:56–64

Martínez ME, De la Barra R, Lobos I, Pavez P (2017) Historia y valorización de la ganadería en Chiloé como diferenciadores de calidad: el caso de la Indicación Geográfica Cordero Chilote. In: I Lobos, Pavez P (eds) Valorización del cordero chilote I.G: Historia, calidad nutritiva y gastronomía de los cortes de carne del primer cordero con denominación de origen en Chile. Osorno, Chile. Instituto de Investigaciones Agropecuaria. Boletín INIA N° 350. Available at https://hdl.handle.net/20.500.14001/6605. Accessed on 16 Feb 2024

Miller S (1980) Human influences on the distribution and abundance of wild Chilean mammals: prehistoric–present. PhD thesis 431 pp. University of Washington, Seattle

Morales R, Subiabre I (2022) Incremento de la grasa infiltrada en carne bovina producida en pastoreo para satisfacer la demanda de carne de alto valor nutricional y calidad sensorial. Osorno, Chile. Instituto de Investigaciones Agropecuarias. Boletín INIA N° 469. Available at https://hdl.handle.net/20.500.14001/68770. Accessed on 16 Feb 2024

Morales R, Folch C, Iraira S, Teuber N, Realini CE (2012) Nutritional quality of beef produced in Chile from different production systems. Chil J Agric Res 72:80–86

Moscoso C, Canto F, Iraira S (2021) Evaluación de disponibilidad y utilización de una pradera naturalizada con y sin fertilización bajo pastoreo ovino: estudio de caso. In: XLVI Congreso Anual Sociedad Chilena de Producción Animal. Chile

Mujica F (2005) Razas ovinas y caprinas en el Instituto de Investigación Agropecuarias. Razas Ovinas. Osorno, Chile. Instituto de Investigaciones Agropecuarias. Boletín INIA N° 127. Available at https://hdl.handle.net/20.500.14001/7027. Accessed on 16 Feb 2024

Oficina de Estudios y Políticas Agrarias (ODEPA) (2013) Estudio: Mercado de la carne y lana ovina en Chile. Informe Final. Available at https://www.odepa.gob.cl/wp-content/uploads/2014/04/mercadoCarneLanaOvina.pdf. Accessed on 16 Feb 2024

Oficina de Estudios y Políticas Agrarias (ODEPA) (2014) Artículo: Carne Ovina. Available at http://www.odepa.gob.cl/wp-content/uploads/2013/10/carneOvina201310.pdf. Accessed on 13 Feb 2024

Peña G, Martínez ME, Calderón C, De la Barra R (2014) Influence of extension on the technological practices and productivity indicators of small-holding sheep farmers in Chiloé, Chile. Int J Agric Exten 2(1):47–55

Petit-Breuilh X, Whitman C, Lagos C, Armijo G, Schiappacasse N (2013) Natural fibre insulation in rural southern Chile. In: PLEA2013—29th Conference, Sustainable Architecture for a Renewable Future, Munich

Ramírez C, Finot S, San Martín C, Ellies A (1991) El valor florístico como indicador ecológico de las malezas del centro-sur de Chile. Agrosur 19(2):94–116

Ramírez-Retamal J, Morales R, Martínez M, De la Barra R (2013) Effect of breed and feeding on the carcass characteristics of Chilote breed lamb. Chil. J. Agric. Res. 73:48–54. https://doi.org/10.4067/S0718-58392013000100007

Ramírez-Retamal J, Morales R, Martínez M, De La Barra R (2014) Effect of the type of pasture on the meat characteristics of Chilote lambs. Food Nutr Sci 5:635–644. https://doi.org/10.4236/fns.2014.57075

Sarricolea P, Herrera-Ossandon M, Meseguer-Ruiz Ó (2017) Climatic regionalisation of continental Chile. J Maps 13(2):66–73. https://doi.org/10.1080/17445647.2016.1259592

Sepúlveda C, Subiabre I, Echeverría R, Morales R (2017) Aceptabilidad, expectativas, preferencias y disposición a pagar de consumidores por cordero chilote I.G. In: I Lobos, Pavez P (eds) Valorización del cordero chilote I.G: Historia, calidad nutritiva y gastronomía de los cortes de carne del primer cordero con denominación de origen en Chile. Osorno, Chile. Instituto de Investigaciones Agropecuarias. Boletín INIA N° 350. Available at https://hdl.handle.net/20.500.14001/6605. Accessed on 13 Feb 2024

Servicio Agrícola y Ganadero (SAG) (2020) Informe Sanidad Animal 2019. Available at https://www.sag.gob.cl/sites/default/files/informe_situacion_sanitaria_animal-2019.pdf. Accessed on 13 Feb 2024

Servicio Agrícola y Ganadero (SAG) (2021) Informe Sanitario Animal 2021. Available online: https://www.sag.gob.cl/sites/default/files/Informe%20Sanitario%20Animal%202021.pdf. Accessed on 13 Feb 2024

Servicio Agrícola y Ganadero (SAG) (2024) Certificación de Productos Orgánicos. Available at https://www.sag.gob.cl/ambitos-de-accion/certificacion-de-productos-organicos/registros. Accessed on 13 Feb 2024

Tadich N, Hernández M (2000) Prevalencia de lesiones pódales en ovinos de 25 explotaciones familiares de la provincia de Valdivia. Chile Arch Med Vet 32(1):63–74. https://doi.org/10.4067/S0301-732X2000000100008

Toro-Mujica P, Riveros JL (2021) Sheep production systems in Chilean Patagonia. Characterization and typology Small Rumin Res 204:106516. https://doi.org/10.1016/j.smallrumres.2021.106516

Chapter 10
Traditional Production System of Iberian Pig and Its Potential for Adaptation to Organic Livestock Farming

Vicente Rodríguez-Estévez, Carolina Reyes-Palomo, Santos Sanz-Fernández, Pablo Rodríguez-Hernández, Ildefonso Caballero-Luna, and Cipriano Díaz-Gaona

Abstract This chapter reviews the traditional production of Iberian pigs and the organization for fattening the pigs by grazing on acorns in the dehesa during autumn-winter. Thus, this system is proposed as a model for producing organic pigs using natural resources and with the lowest carbon footprint, along with other environmental benefits that livestock production in agroforestry systems offers. Details of the organization of the farms and the requirements for conversion to organic production are also provided.

Keywords Acorns · Agroforestry · Carbon footorint · Dehesa · Ecosystem services · Fatty pig · Free-range · Grazing · Iberian breed · Sustainability

10.1 Introduction

The Iberian pig is a native breed raised in the southwest of the Iberian Peninsula (Spain and Portugal), in the dehesa agroecosystem (called "montado" in Portugal). The Spanish Society for the Study of Pastures (S.E.E.P.) defines "dehesas"as surfaces with trees that are dispersed and a well-developed herbaceous stratum, with the stratum of shrubs largely eliminated. These areas have an agricultural origin (ploughed land in long-term rotations) and are used mainly for extensive or semi-extensive grazing, foraging grasses, browse pastures, and fruits of trees (Ferrer et al. 1997). This landscape resembles a savanna but with trees of the genus Quercus.

V. Rodríguez-Estévez (✉) · C. Reyes-Palomo · S. Sanz-Fernández · P. Rodríguez-Hernández · I. Caballero-Luna · C. Díaz-Gaona
Cátedra de Producción Ecológica Ecovalia-Clemente Mata, Dpto. de Producción Animal, Universidad de Córdoba, Córdoba, Spain
e-mail: pa2roesv@uco.es

C. Palacios Riocerezo, M. Gonzalez Ronquillo (eds.), *Traditional Livestock Production*, https://doi.org/10.1007/978-3-031-88275-3_10

Costa et al. (2006) indicate that the evergreen oak (Q. ilex rotundifolia) is the predominant species, covering 70.1% of the dehesa surface. The production of Iberian pig is a rare case in the world of pork production where the pig farming collaborates significantly in the preservation of the ecosystem (Lopez-Bote 1998; Rodríguez-Estévez et al. 2012).

The Iberian breed shows lower production performance in traditional extensive systems compared to intensive systems; for instance, farrowing interval can be more than one month longer because extensive sows lactation lasts from 45 to 60 days. The lower production pressure of these extensive animals, combined with freedom of movement and diet based on natural sources grazed (mainly acorns and grass for finishing during winter), offer these animals a higher level of welfare than intensive farms. This breed has its own regulation concerning commercial labeling and quality control, which basically refers to racial purity, feeding regimen, weights, and ages for slaughter (Ministerio de Agricultura, Alimentación y Medio Ambiente 2014).

The traditional finishing of Iberian pigs is based on a free-range system where they forage for acorns, aligning with the principles of Agroecology for Organic Agriculture (which prioritize health, ecology, equity, and responsibility), as well as with current European regulations for organic production (European Parliament and Council 2018. In this sense, the traditional Iberian pig farming system is in a very advantageous position to undertake its organic conversion and thereby differentiate itself in a European market that increasingly values animal welfare and sustainability in production systems. This chapter reviews traditional Iberian pig farming as a model for adaptation to organic livestock farming.

10.2 Characteristics of the Iberian Pig Breed

The Iberian pig comes from the Mediterranean pig lineage (*Sus mediterraneus* o *Sus scrofa meridionalis*) and it has evolved in dehesa agroecosystem. This is one of the last grazing pig breeds in Europe, thanks to its great rusticity and ability to exploit the resources of the dehesa, even in times of scarcity and under adverse climatic conditions. The dehesa grasslands are annual herbaceous pastures, characterized by semi-arid climate (Mediterranean) and strong seasonality with a long dry season (frequently from June to October, both months included). Faced with these circumstances, Iberian pigs present a series of adaptive advantages that make it the most efficient animal in transforming the resources of this ecosystem, withstanding periods of summer scarcity to quickly fatten with autumnal acorn mast (*montanera* in Spanish, meaning pannage). Therefore, these pigs have been traditionally slaughtered only at a specific time of the year, from the end of December to February–March. By other side, the sturdy limbs, along with the pigmentation of its skin, allow the Iberian pig the adaptation to rugged terrain and high temperatures (Lopez-Bote 1998).

Table 10.1 Productive data of the main varieties of the Iberian pig

	Lampiño	Entrepelado	Retinto	Torbiscal
Carcass weight (kg)	159.40	143.20	139.40	157.50
Carcass yield (%)	86.87	83.78	83.63	82.66
Hams vs carcass (%)	14.00	15.00	15.00	16.00
Shoulders vs carcass (%)	9.00	10.00	10.00	10.00
Prime cuts vs carcass (%)	23.00	25.00	25.00	26.00

Sources: Delgado Bermejo et al. (2001), Clemente-López et al. (2007)

The Iberian pig is often considered as a racial grouping made up of different lines or varieties. Their skin is always pigmented, ranging from black varieties to blondes or red-tinged, with sparse and weak hair (*Entrepelado* variety) or absent (*Lampiño* variety). They have an approximate weight of 100–150 kg for no pregnant females, with a series of characteristic physical traits: they are pigs with elongated snouts, ears folded towards the face, very developed dewlaps in fattening animals and short, fine but resistant limbs, ending in characteristic black-coloured hooves (Clemente-López et al. 2007), except for the scarce variety *Torbiscal* which can present depigmented or whitish-striped hooves.

Nieto et al. (2019), reviewing about this breed, indicate that Iberian pigs are slaughtered at approximate age of 407 days and reached around 152 kg live body weight, with dressing yield around 81%, and 85 mm average back fat thickness in studies with final body weight above 100 kg. Some carcass yield data of the Iberian pig main varieties are available in Table 10.1. At the end of the traditional finishing phase (140–160 kg BW), based on foraging acorns, they can reach 60% carcass fat, 15 cm backfat thickness and 10–13% intramuscular fat content (López-Bote 1998). The quality of the meat of those pigs is emphasized; mainly due to its very specific properties and healthy mono-unsaturated fats with a high content of oleic acid (around 55%) from acorn diet and a very low concentration of linoleic and palmitic acids (around 8 and 20% respectively) (Flores et al. 1988).

Unlike other pig breeds, the Iberian pig has a low susceptibility to stress, a high tendency to accumulate fat, as well as a high level of fat marbling in the muscles, this being a key characteristic that determines the quality of the pork and cured products (Lopez-Bote 1998). Besides that, there are different factors that decisively influence this fat infiltration and thus the final quality of the products. Regarding breed purity, it has been shown that the level of infiltration and the regular marbling are slightly higher in purebred Iberian pigs than in crossbred pigs, where such infiltrations are larger and less evenly distributed (Fernández et al. 2003; Tejeda et al. 2002). Nutrition also influences the level of fattening and the type of intramuscular fat, being of greater quantity and quality in animals fed exclusively on acorns (Cava et al. 1997; Tejeda et al. 2002). Lastly, the level of exercise of the animals, which determines the concentration of myoglobin in their muscle fibers, is also related to the fat infiltration in the meat. Therefore, higher quality products come from animals that are fattened by foraging on acorns. These significant development of subcutaneous fat and abundant intramuscular infiltration, allow for the curing and maturation of meat products to occur optimally in a dry and hot climate.

More information about this breed can be found in the review by Nieto et al. (2019).

10.3 Traditional Management Practices

10.3.1 Organization of Batches and Farrowings

Gilts have their first mating at approximately 10 months old and sows can wean 7–8 piglets per litter. The traditional reproductive scheme of the Iberian sow is organized to achieve 2 parturitions per sow per year, with all the sows in the herd farrowing on the same dates, following a reproductive cycle based on lactations that can exceed 45 days. Later, all sows are weaned at the same time and grouped to be mated by a group of boars through natural mating, with which they remain for a month, allowing for a second mating for repeated sows. Thus, all sows can potentially have two cycles of gestation and lactation in the same year, occupying farrowing crates or huts twice a year for about two months each time, with birthing dates typically around December and June (Fig. 10.1).

The breeding dates are aimed at ensuring that farrowings and lactations coincide with favourable seasons, and the entire subsequent life of the growing pigs is organized so that they reach a suitable weight at the beginning of the *montanera* in autumn. This allows for their fattening to be based on foraging, taking advantage of the compensatory growth that follows a previous stage of slow growth with dietary restriction. To better utilize the farrowing facilities, some farmers divide the herd into 2 batches and alternate farrowing every three months (December, March, June, and September) (Fig. 10.2). As the prolificacy of these sows is low, with this system, their annual productivity usually does not exceed 13–14 piglets weaned per sow per year.

Cycles/year	1st cycle						2nd cycle					
Months	1	2	3	4	5	6	7	8	9	10	11	12
Mating (1 month/cycle)	X											
Gestation (113 days/cycle)		X	X	X	X							
Lactation (from 45 to 60 days/cycle)						X						
Mating (1 month/cycle)							X					
Gestation (113 days/cycle)								X	X	X	X	
Lactation (from 45 to 60 days/cycle)												X

Fig. 10.1 Traditional reproductive scheme of two batches of Iberian sows for two yearly parturitions per sow

Fig. 10.2 Traditional farrowings' calendar with one and two flocks of sows for finishing pigs doraging acorns in Autumn-Winter

10.3.2 Seasonal Variations in Diet and Food Resources

Iberian piglets were traditionally weaned at around 8 weeks old, although currently it is more common to wean them at around 40–45 days old. After weaning, the management of the piglets varies depending on their birth season, but the entire traditional production cycle of Iberian pigs was structured to ensure their physiological readiness for foraging acorns during the finishing phase, known as the *montanera*. An integral component of their traditional management involves an extended period of growth (or pre-fattening) and controlled feeding, with a diet derived from natural resources (adjusted according to the available resources in each dehesa): spring grasses, summer stubble, agricultural by-products, etc. This approach aims to capitalize on the pigs' compensatory growth, as the objective is for them to reach a weight of 90–110 kg by October, so they can be finished during the *montanera* period in the autumn and winter months. Therefore, there are several options:

(a) When piglets are born between November and January, they typically undergo their first post-weaning growth stage during spring grazing until they reach a weight of 50–55 kg. The second growth stage, prior to being fattened on acorns, occurs from June to October. During this period, they graze on stubble fields and the limited dehesa pastures available in summer, along with any supplementary feed offered, to reach a weight of around 100 kg and an adequate skeletal development by late October. The feed rationing in this phase is around 3% of the live weight.

(b) When piglets are born between February and April, they are either raised with a longer cycle to take advantage of the following year's *montanera*, finishing their fattening around 24 months old, or with a shorter cycle, utilizing part of the *montanera* remains from the year of birth and being finished with feed and spring pastures. However, these pigs cannot be marketed as acorn-finished pigs.

(c) When piglets are born between May and July, their growth was traditionally slowed down so that they did not reach a suitable weight for starting the *montanera* until their second autumn. Nowadays, these pigs are typically finished with a diet primarily composed of feed, although they may forage on some resources, but they do not have access to forage acorns.

(d) When piglets are born between August and October, currently their growth is also traditionally slowed down to reach a suitable weight for starting the *montanera* at their second autumn, which is the traditional finishing of this farrowing flock.

Ethnobotanical reviews can inspire the use of local minor crops, herbs, and forages from various sources for pig feed. The European project OKnet Ecofeed (https://ok-net-ecofeed.eu/) showed that in the Iberian Peninsula, over 140 species have been used as forage or feed for pigs (Rodríguez-Estévez et al. 2022a). The evergreen oak (*Quercus rotundifolia*) and the cork oak (*Q. suber*), with their acorns, are just two examples, though they are particularly important resources that underpin the entire traditional Iberian pig production system in the dehesa. Therefore, traditional livestock systems and the resources they utilized deserve attention for their potential as sources of inspiration for innovation. These systems and resources can be adapted to current social and market circumstances. It is important to note that raising pigs for family subsistence farming will always be more versatile and adaptable compared to professional farming, which requires careful planning and proper diet formulation (Rodríguez-Estévez et al. 2022b).

In this regard, Rodríguez-Estévez and Mata-Moreno (2006) describe the traditional diet calendar for Iberian pigs in the Sierra de Cádiz district, which relied mainly on natural resources foraged and agricultural byproducts (Fig. 10.3). It's worth noting that only at the end of summer were some grains (barley and broad or fava beans) incorporated into the pigs' diet, although those production cycles were very long, as previously mentioned. Acorns of different oaks species are the main diet resource in autumn and winter in that calendar; like that indicates, many medieval calendars depict men beating down acorns to feed their pigs on the page and illustration for November (Fig. 10.4).

The use of local resources to feed pigs has traditionally included by-products, forages, and fruits, leveraging the rusticity and adaptability of this native breed. Pigs have been traditionally managed as the livestock species that recycles food resources within the agricultural system. So, farrowing dates are scheduled to ensure that the fattening period coincides with times when the food resource used as the energy source for fattening is most abundant.

The system of fattening pigs on forest fruits, as occurs during the *montanera* (pannage) with acorns (mainly *Quercus rotundifolia*), can be extended to other fruits such as chestnuts (*Castanea sativa*), beechnuts (*Fagus sylvatica*), or carob beans (*Ceratonia siliqua*), as well as any horticultural byproduct from discards or

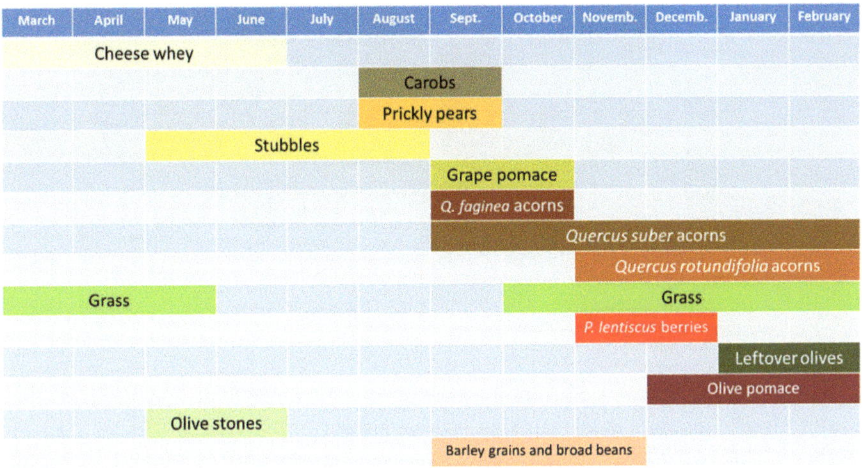

Fig. 10.3 Example of traditional diet calendar for Iberian pigs based on natural resources foraged and agricultural byproducts

Fig. 10.4 Examples of medieval calendars depicting men beating down acorns to feed their pigs on the page for November

surpluses. Olive by-products, such as olive pomace, or even unharvested olives, are another valuable resource for feeding Iberian pigs. These last olives can be foraged fresh during winter, naturally dried in mid-spring, or consumed as seeds after cracking the olive stones in late spring (Fig. 10.5), taking advantage of their high quality and fat content.

Fig. 10.5 Dried unharvested olives and Iberian pigs foraging and cracking their stones in late spring to consume their seeds

Other food resources traditionally used as pig fodder include forage crops like alfalfa (*Medicago sativa*) or sainfoin (*Onobrychis viciifolia*), tree forages such as mulberry (*Morus alba* and *Morus nigra*), local varieties of legumes like lupine (*Lupinus* sp.), chickpea (*Cicer arietinum*), beans (*Phaseolus vulgaris*), or peas (*Pisum sativum*), and stubbles from cereal crops (Rodríguez-Estévez et al. 2022a, b).

In this regard, the European Commission (2018) published the "Guidelines for the feed use of food no longer intended for human consumption," which opens new perspectives on using food by-products in line with the circular economy aim of organic production. This could also reduce the land needed for pig food production in Europe (Zu Ermgassen et al. 2016).

10.3.3 Free-range Grazing and Acorn-Based Feeding

The finishing phase in *montanera* (pannage) does not commence until there is an adequate supply of acorns for pigs to forage and to fatten just grazing these and other natural resources, but mainly weeds. During this interim period, pigs are fed in a restricted manner. The beginning of *montanera* season is influenced by weather conditions (specifically, the first autumn rains for grass growth) and the availability of ripe acorns (mast) on the ground, which the pigs forage from October to February. According to the meta-analysis data by Sánchez-Esquiliche et al. (2023), at the beginning, pigs weighing 100 kg are between 250 and 425 days old, while at the end, pigs weighing 150 kg ranged from 350 to 525 days old. This indicates that the growth rate of free-range fattened Iberian pigs is not solely dependent on age, but also on the preceding restrictive feeding regimen to take advantage of this species compensatory growth. The age at which pigs begin this finishing period is not fixed and is determined by the availability of acorns on the ground; but usually these pigs are older than 12 months.

Studies, based on direct and continuous on-site observations of ingestive bites taken by pigs under continuous monitoring during foraging, demonstrate that the Iberian pig *montanera* diet primarily consists of acorns and grass, accounting for

Fig. 10.6 Finishing Iberian pigs foraging acorns in the dehesa

56.5% and 43.3% of grazing bites, respectively. Additionally, only nine other resources (such as berries, bushes, roots, carrion, straw, etc.) were consumed at a frequency ≥0.01% because pigs behave as opportunistic omnivorous species (Rodríguez-Estévez et al. 2009a). This translates to a daily intake of 1251 to 1469 acorns or 7.13 to 8.37 kg of whole acorn, and 2 to 2.7 kg of grass, during 6.1 to 7.1 hours of foraging (Fig. 10.6). With that intake the average daily weight gain of Iberian pigs has been found to be 0.79 ± 0.03 kg during *montanera* fattening period. So, the corresponding food conversion rate, expressed in terms of whole acorns required to achieve the reported growth rate, taking into account the contribution of grass, is 10.5 ± 0.75 (Rodríguez-Estévez et al. 2010). This is the key figure to determine the carrying capacity for free-range finishing pigs foraging on acorns.

Rodríguez-Estévez et al. (2011) studied the average daily weight gain (ADWG) of two flocks of Iberian pigs during the first two months (from November 1st) of two consecutive *montaneras* on the same dehesa farm. They found a mean ADWG of 0.76 ± 0.01 kg/day; however, variations were observed based on the year (0.74 ± 0.02 vs. 0.78 ± 0.02 kg/day, $p < 0.05$), sex (0.78 ± 0.02 vs. 0.72 ± 0.02 kg/day for males and females, respectively; $p < 0.01$), and age (0.71 ± 0.02 vs. 0.83 ± 0.02 kg/day for pigs younger than a year and older than a year at the beginning, respectively, $p < 0.001$). Therefore, considering these factors is important for planning the slaughter weight and age or date.

In this regard, considering the age and weight of the pigs at the start of the finishing period, Sánchez-Esquiliche et al. (2023) conducted a meta-analysis and developed a two-variable model to predict the ADWG during this finishing period:

$$f(x,y) = -0.6199 + 0.1135x - 0.031y - 0.0002x^2 - 0.0001xy + 6.85 \cdot 10 - 5y^2$$
$$(\text{root mean square error or RMSE} = 0.1249);\ \text{where } x = \text{age, and } y = \text{weight.}$$

This mathematical model allows for the estimation of ADWG and the determination of the date or age at which these pigs could be sent to the slaughterhouse, as well as their target weight.

10.4 Sustainability of Traditional Iberian Pig Farming

10.4.1 Stocking Rates Calculation

Regardless of the established upper limits of pig density to prevent contamination from excreta (<170 kg N/ha) (Council of the European Communities 1991), stocking rates exceeding 4 pigs/ha throughout the year pose a risk of deterioration of dehesa pastures and soil due to the behavior of pigs. In the pigs' outdoor runs, where they are free to wander but do not have forage to graze, there is a high risk of soil deterioration, contamination of water (aquifers and nearby watercourses) both by infiltration and runoff of excreta, and tree mortality. Therefore, in areas with high pig concentration, if rotational presence is not possible, it would be preferable to have a smaller, concreted, and well-conditioned area for easy cleaning and effluent collection.

In compliance with the current quality standard for the Iberian pig (Ministerio de Agricultura, Alimentación y Medio Ambiente 2014), it is mandated that they attain the slaughter weight (≥161 kg) solely through the consumption of natural resources (mainly acorns and grass) obtained during grazing in autumn and winter. Understanding the high food conversion rate of finishing Iberian pigs in the dehesa (10.5 ± 0.75 kg of whole acorns to gain 1 kg, in addition to grass intake) is crucial for determining their stocking rate during the montanera season (Rodríguez-Estévez et al. 2010). Therefore, considering that an adult evergreen oak (*Q. ilex rotundifolia*) yields an average mast of 11 kg of acorns (Rodríguez-Estévez et al. 2007), it could be inferred that a grazing Iberian pig necessitates the entire annual acorn production of an adult evergreen oak to achieve a 1 kg weight gain (Rodríguez-Estévez et al. 2010). Consequently, the stocking rate can be estimated by dividing the number of adult oaks in a dehesa state by the expected weight gain; a minimum of 46 kg according to quality standards (Ministerio de Agricultura, Alimentación y Medio Ambiente 2014). Moreover, considering the selective feeding behaviour of Iberian pigs towards acorns with preferred characteristics (Rodríguez-Estévez et al. 2009b), the aforementioned quotient should be viewed as a minimum requirement to ensure finishing based solely on acorns and grass.

Therefore, considering that a dehesa typically has an average of 30–40 mature trees per hectare (Fig. 10.7), an Iberian pig will require more than one hectare of dehesa to complete its fattening solely on acorns. What is more, dehesa oaks (*Quercus* sp.) have a strong tendency to alternate bearing or alternation in acorn production or mast.

10.4.2 Ecosystem Services of Free-Range Iberian Pig Production System

It has been already indicated that the traditional Iberian pig pasture-based system is mainly located in the dehesa agroecosystem (Fig. 10.7), which is a High Natural Value System (Plieninger et al. 2021). It is included in the EU Habitats Directive as

Fig. 10.7 Dehesa agroecosystem and free-range growers grazing during spring to be prepared to start their finishing foraging acorns six months later

habitat code 6310 (Council of the European Communities 1992). This directive considers it as a "natural habitat type of community interest" included in the "natural and semi-natural grassland formations", where it is called "sclerophyllous grazed forests (dehesas) with *Quercus suber* and/or *Q. ilex*"; besides, it advises the designation of special areas for dehesa conservation.

Extensive livestock systems contribute to generate a great range of ecosystem services, such as prevention of forest fires, the production of specific quality products linked to the territory, biodiversity conservation and cultural landscapes (Aguilera et al. 2020; Díaz-Gaona et al. 2014).

Biodiversity conservation is one of the main ecosystem services provided by the extensive livestock system based on the dehesa. This is home to wildlife typical of Mediterranean forests and is further enriched by species from other habitats such as steppes and agricultural areas. It is widely acknowledged for its exceptional conservation value. About 30% of the Iberian Peninsula's vascular plant species are found in the dehesas (Pineda and Montalvo 1995). Marañón (1985) identified 135 species in a 0.1 ha plot in an Andalusian dehesa, considering it one of the most diverse vegetation types in the world at this scale, particularly among Mediterranean ecosystems (Fig. 10.8). Dehesas also serve as a habitat for several rare or globally threatened species, including black vultures (*Aegipius monachus*), Spanish imperial eagles (*Aquila adalberti*), and Iberian lynx (*Lynx pardinus*).

Furthermore, the principle of farm diversification emerges as a key strategy, particularly evident in traditional and extensive livestock production systems (Plieninger et al. 2006) (Fig. 10.8 left). In the Iberian pig productive systems, the main strategy is its combination with ruminant production, but alternative income diversification by integrating multiple agricultural and non-agricultural activities can be found.

Climatic stability is another important ecosystem service provided by the dehesa extensive livestock systems; and these also offer potential for boosting soil carbon accumulation (Soussana et al. 2004). Specially in Mediterranean and semi-arid pastures, grazing has been shown to enhance belowground carbon allocation (Whitehead 2020). However, stocking rates (SR) must align with pasture productivity to avoid

Fig. 10.8 Left: pregnant Iberian sows grazing in a dehesa during spring in Natural Park Sierra de Grazalema (Villaluenga del Rosario, Cádiz, Spain); right: lactating Iberian sows and piglets and cattle in Los Pedroches (Villanueva de Córdoba, Spain)

decreasing soil organic carbon (SOC) stocks and hindering tree recruitment (Abdalla et al. 2018), because high stocking rates can diminish soil fertility (Fernandez et al. 2003), reduce grassland's CO_2 capture potential by limiting aerial biomass (Frank et al. 2002), and increase soil compaction (Byrnes et al. 2018; Garnett et al. 2017).

When assessing the carbon footprint of the extensive production of Iberian pigs its result is clearly favourable. This is because, besides generating lower greenhouse gas emissions due to the absence of animal feeds consumption in its final production stage, carbon sequestration is promoted in soil, pastures, and trees, leading to a negative carbon footprint. Reyes-Palomo et al. (2023) calculated their carbon footprint with a "cradle to farm gate" approach using kg of live weight (kgLW) as functional unit, and including C sequestration in soils and biomass. The average carbon footprint of the final product was -5.6 ± 6.5 kg CO_2eq kgLW^{-1} for free-range fattened pigs (finished grazing acorns) (Fig. 10.9).

The main C sinks of the dehesa system (trees and soil) sequestered an average of 3.18 t CO_2eq ha-1 year-1. This C sequestration, averaging -10.2 ± 6.7 kg CO_2eq kgLW^{-1} in free-range pig production, offsets 223% of the GHG emissions in those systems, which resulted in a negative C footprint (net CO_2 sequestration). What is more, there was a direct relationship between CF and pig stocking rate, leading to strongly negative CF values on those farms with the lowest pig stocking rates (Reyes-Palomo et al. 2023).

This fact for climate change mitigation is enhanced by organic production that further reinforces these values (organic feed has a lower C footprint), could serve to boost a system that offers many environmental benefits for society.

Moreover, beyond environmental considerations, the ethical dimensions of livestock production cannot be overlooked, especially given evolving consumer preferences towards sustainability and animal welfare. Extensive farming practices inherently align with ethical production standards by allowing animals to express natural behaviour patterns (Pietrosemoli and Tang 2020) enhancing overall animal welfare.

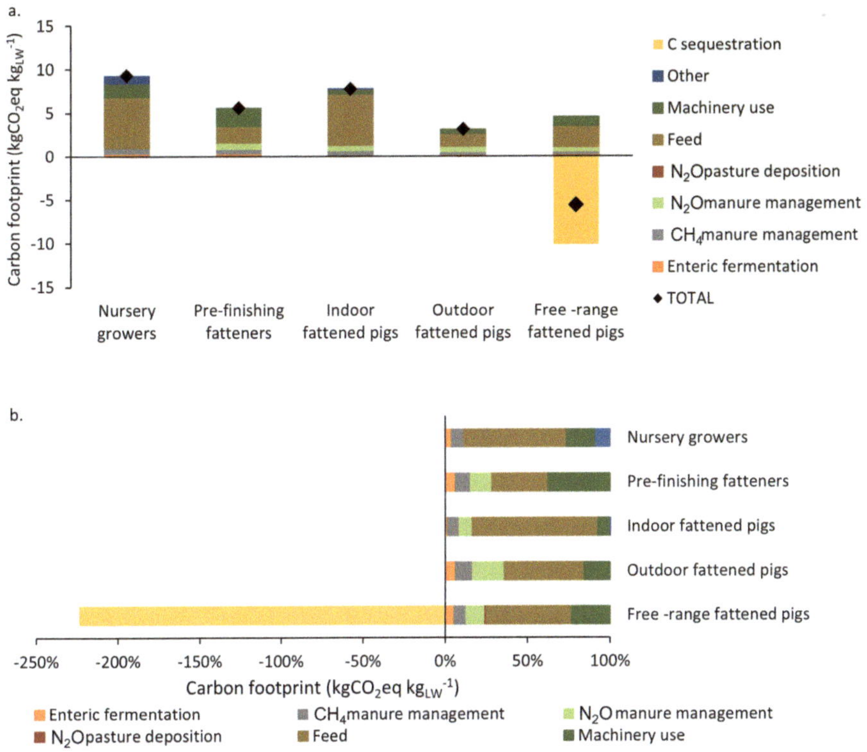

Fig. 10.9 Carbon footprint (kg CO_2 eq kgLW-1) of the different categories of conventional Iberian pig productions on dehesa farms showing the different sources of emissions and carbon sequestration. A. Absolute value; b. Relative values (in percentage) (Reyes-Palomo et al. 2023), (*LW* live weight)

10.4.3 Challenges and Limitations of Traditional Production Methods

Extensive systems are not only located in less productive areas, but they are also less efficient and less productive than the intensive systems, and according to some authors, the most extensive systems are also the less productive ones (Escribano et al. 2016). This lack of productive capacity or efficiency of the most extensive systems lead to a lower economic performance, favouring more efficient and intensive systems, which bring greater economic benefits.

Those limitations are well represented in the Iberian pig productive system, where extensively raised Iberian sows wean around 12 piglets per sow per year, compared to intensively raised sows that wean an average of 14.78 piglets per sow per year. In contrast, commercial modern breeds wean an average of 29.38 piglets per sow per year (Sanz-Fernández et al. 2024). At the same time, these animals take longer to finish than commercial breeds due to their productive cycle, which increases the economic effort and management dedication required to achieve the specified weight at the appropriate times. The goal is to finish them during the *montanera* period, fattening them while making the best use of the foraged acorn mast. However, there is a strong tendency for alternate bearing or alternation in acorn mast production, pigs are born more than a year before without any prediction of acorn mast availability.

10.5 Transitioning to Organic Livestock Farming

10.5.1 General Requirements for Organic Farming

Applying the principles and requirements established in EU Regulation 2018/848, two models of organic pig farms could be implemented: closed-cycle farms or farms specialized in certain phases (reproduction/rearing/fattening) where animals born from organic sows are raised. In both cases, although not mandatory, batch management is recommended to rationalize management, better meet the nutritional needs of the animals, and ensure the handling hygiene principle of all in-all out.

It is required that animals have access to shelters, and shade is especially important in Mediterranean climate zones during hot periods. Animals must have the maximum possible space, always above the minimum established in Chap. II of Implementing Regulation No 2020/464, with earth or concrete yards and buildings with continuous floors, not slatted. Good environmental conditions (light, ventilation, insulation, etc.) must exist within the housing de facilities, with straw bedding or other manipulable materials that must also be available in the outdoor runs, unless these are pasturelands.

Breeding sows may have huts or any other shelter for raising their piglets. Farrowing pens are allowed if their surface is at least 10 m^2/sow (minimums of 7.5 m^2 indoors and 2.5 m^2 outdoors of pen or yard). Nonbreeding or gestating sows must be housed in groups with at least 4.4 m^2/sow (minimums of 2.5 m^2 indoors and 1.9 m^2 outdoors of pen or yard). During postweaning phase, the minimum surface allowance must be 1 m^2, always with outdoors access. In the case of traditional Iberian pig systems these areas are far exceeded.

Regarding feeding, farms themselves must produce at least 30% of what the animals consume, which is widely met during the fattening phase foraging acorns and grass (*montanera* or pannage) with fewer than 1 pig/ha. In other cases, farms must have a crop area to obtain a portion of the grains consumed by the animals. This cultivation area can be owned or belong to farmers with whom a long-term agreement is maintained, where part of the composted manure or bedding from the farm will normally be recycled. In more intensive systems, although this is not the case in traditional Iberian pig production, forages must be provided in accordance with the standards' regulation; which, in addition to food, provide entertainment for the animals and contribute to their welfare. Besides that, the piglets' lactation length must be at least 40 days.

Regarding disease control, this is based on prevention through ensuring optimal welfare conditions, a high level of biosecurity and hygiene measures (considering outdoors limitations) based in paddocks and pastures rotations. Since vaccines are allowed, the implementation of a vaccination plan is essential. One important contagious disease to consider in vaccination programs is swine erysipelas, caused by *Erysipelothrix rhusiopathiae* and seen mainly in growing pigs and characterised clinically by sudden death, fever, skin lesions and arthritis. In the case of using synthetic chemical medicines, the withdrawal period is double the established suppression time or at least 48 h when it is zero. Group treatments without prior diagnosis (chemoprevention) are prohibited, including deworming.

The castration of animals, a traditional practice aimed at improving the organoleptic quality of products, can be carried out following veterinary regulations either by surgery or by applying Improvac®, an immunological castration product for pigs. Improvac® reduces the production of hormones responsible for boar taint in pork, offering an alternative to surgical castration. Castration is also important to prevent the attraction of wild boars and the associated health risks.

10.5.2 Potential Challenges and Opportunities for Transitioning Traditional Iberian Pig Farming to Organic Practices

Firstly, the transition or conversion of traditional Iberian pig farms to organic production is an opportunity to improve their sustainability and profitability. However, marketing organic Iberian pork products has, as its main drawback, their high price, as this is the case even with non-organic ones. However, since the differences with the latter are not significant, more and more companies are dedicating themselves to it. Some are family businesses that exclusively work with organic products; other farmers work with conventional Iberian pork and subsequently, to diversify their product offerings, have opened an authorized organic Iberian pig farm to offer the highest quality with the greatest guarantees to the most demanding consumers.

Effectively communicating to customers, the distinctive features of the products and the guarantees of organic production system is crucial, especially regarding the ecosystem services associated with this production system (landscape and biodiversity conservation, CO_2 sequestration, etc.). It's also about ensuring maximum respect for the environment and animal welfare throughout all breeding phases, not just during the *montanera*.

Secondly, it is important to aim for moderate production based on the optimal use of natural resources and stocking rates, with acorn mast as the main finishing food resource. It is crucial to avoid forcing productivity, as this requires intensive management and high investment in resources. Traditional and organic production is compatible with profitability, which can be achieved through high productive efficiency and a balance between the resources invested across the entire farm. This essentially means maintaining adequate stocking rates in balance with carrying capacity and acorn mast forecasts.

Regarding regulatory requirements, the main challenge is probably ensuring 100% organic feeding throughout all breeding phases, and especially achieving a minimum of 30% self-sufficiency in food. This can be achieved through partnerships with nearby farmers who start the transition simultaneously with the pig farm. And by utilizing the cultivable land of the farm itself, which in many cases has been abandoned or underutilized for years. The best dehesa soils can be ploughed and grown every 3–4, or more years to obtain hay and even some grains (Fig. 10.10). Additionally, it is necessary to work with balanced diets and have information on what pastures and different products and by-products contribute to the pigs' diet. The quality of Iberian pork in organic farming will directly depend on their feeding and must meet consumer quality demands.

In this regard, in organic farming, feed management focuses on maximizing natural resources while minimizing reliance on external inputs. That is an important factor to reduce farming C footprint. Moreover, several life cycle assessment (LCA) studies about pig production (McAuliffe et al. 2016; Nguyen et al. 2011; Reyes-Palomo et al. 2023) indicate that feed consumption associated emissions are the greatest source of GHG emissions in all the productive phases, regardless of the final fattening model.

Furthermore, facilities for animals and disease control do not pose major challenges. However, when sows are in free-stall housing with small outdoor runs, aiming for higher productivity (above 2 farrowings per sow per year and more than 7 weaned piglets per farrowing, or farms producing batches weekly or every three weeks), management can become more complicated. In any case, basic hygiene principles, such as all-in-all-out and outdoor run rotations, are key. Moreover, the

Fig. 10.10 Iberian pig growers (Torbiscal variety) foraging in cereal stubble field and unharvested sowing (pigs of Mío1898® farm)

traditional system of two parturitions per year for the entire flock ensures around four months of sanitary break between different batches in the farrowing huts and other facilities.

10.6 Conclusion and Future Directions

In essence, this traditional free-range pig farming system, characterized by its emphasis on adaptation to local resources and environments, emerges as a resilient solution in navigating the complexities of contemporary socio-economic landscapes. By fostering self-sufficiency, promoting diversification, upholding ethical standards, driving rural development, and ensuring food security, extensive farming systems not only sustain livelihoods but also contribute to the broader goals of sustainability and resilience in agricultural practices.

Market opportunities for organic pork products are on the rise due to the growing demand for healthy, ethically produced, and environmentally friendly food. Today, many consumers are willing to pay relatively high premiums for natural and organic pork, and producers who can meet this demand could capitalize on a growing market and generate higher revenues.

In conclusion, practical considerations for organic Iberian pig farming encompass a range of strategies aimed at promoting sustainability, ensuring regulatory compliance, and maintaining product quality. By prioritizing organic feed management, adhering to certification requirements, and implementing robust labelling practices, farmers can uphold the integrity of traditional production systems while meeting the evolving demands of organic agriculture.

References

Abdalla M, Hastings A, Chadwick DR, Jones DL, Evans CD, Jones MB, Rees RM, Smith P (2018) Critical review of the impacts of grazing intensity on soil organic carbon storage and other soil quality indicators in extensively managed grasslands. Agric Ecosyst Environ 253:62–81. https://doi.org/10.1016/j.agee.2017.10.023

Aguilera E, Díaz-Gaona C, García-Laureano R, Reyes-Palomo C, Guzmán GI, Ortolani L, Sánchez-Rodríguez M, Rodríguez-Estévez V (2020) Agroecology for adaptation to climate change and resource depletion in the Mediterranean region. A review. Agric. Syst. https://doi.org/10.1016/j.agsy.2020.102809

Byrnes RC, Eastburn DJ, Tate KW, Roche LM (2018) A global meta-analysis of grazing impacts on soil health indicators. J Environ Qual 47:758–765. https://doi.org/10.2134/jeq2017.08.0313

Cava R, Ruiz J, López-Bote C, Martín L, García C, Ventanas J, Antequera T (1997) Influence of finishing diet on fatty acid profiles of intramuscular lipids, triglycerides and phospholipids in muscles of the Iberian pig. Meat Sci 45:263–270. https://doi.org/10.1016/S0309-1740(96)00102-7

Clemente-López I, Diéguez-Garbayo E, Forero-Vizcaíno FJ (2007) Las razas porcinas autóctonas andaluzas: El Cerdo Ibérico y sus estirpes. El Manchado de Jabugo, in: Patrimonio Ganadero Andaluz: Las Razas Ganaderas de Andalucía (Volumen II). Consejería de Agricultura y Pesca, pp. 367–410

Costa JC, Martín A, Fernández R, Estirado M (2006) Dehesas de Andalucía: caracterización ambiental. Consejería de Medio Ambiente de la Junta de Andalucía, ISBN: 84-96329-81-X, Sevilla

Council of the European Communities (1991) Council Directive of 12 December 1991 concerning the protection of waters against pollution caused by nitrates from agricultural sources (91/676/EEC). Off J Eu Commun L 375:1–8

Council of the European Communities (1992) Council Directive 92/43/EEC of 21 May 1992 on the conservation of natural habitats and of wild fauna and flora. Off J Eu Commun L 206:7–50

Delgado Bermejo JV, Diéguez Garbayo E, Cañuelo Rojas P, Barba Capote C (2001) Caracterización productiva de las variedades del cerdo ibérico basada en el comportamiento en matadero. Arch Zootec 50:125–132

Escribano AJ, Gaspar P, Mesías FJ, Escribano M (2016) The role of the level of intensification, productive orientation and self-reliance in extensive beef cattle farms. Livest Sci 193:8–19. https://doi.org/10.1016/j.livsci.2016.09.006

European Parliament and Council (2018) Regulation (EU) 2018/848 of the European Parliament and of the Council of 30 May 2018 on organic production and labelling of organic products and repealing Council Regulation (EC) No 834/2007. Off J Eu Union L 150:1–92. http://data.europa.eu/eli/reg/2018/848/oj

Fernández A, De Pedro E, Núñez N, Silió L, García-Casco J, Rodríguez C (2003) Genetic parameters for meat and fat quality and carcass composition traits in Iberian pigs. Meat Sci 64:405–410. https://doi.org/10.1016/S0309-1740(02)00207-3

Ferrer C, San Miguel A, Ocaña M (1997) Propuesta para un nomenclátor definitivo de pastos en España. Pastos 17(2):125–161

Flores J, Birón C, Izquierdo L C, Nieto P (1988) Characterization of green hams from Iberian pigs by fast analysis of subcutaneous fat. Meat Sci 23:253–262

Frank DA, Kuns MM, Guido DR (2002) Consumer control of grassland plant production. Ecology 83:602–606. https://doi.org/10.2307/3071865

Garnett T, Godde C, Muller A, Röös E, Smith P, De Boer IJM, Zu Ermgassen E, Herrero M, Van Middelaar CE, Schader C (2017) Grazed and confused?: ruminating on cattle, grazing systems, methane, nitrous oxide, the soil carbon sequestration question-and what it all means for greenhouse gas emissions. FCRN

Lopez-Bote CJ (1998) Sustained utilization of the Iberian pig breed. Meat Sci 49:S17–S27. https://doi.org/10.1016/S0309-1740(98)90036-5

Marañón T (1985) Diversidad florística y heterogeneidad ambiental en una dehesa de Sierra Morena. Anales de Edafología y Agrobiología 44:1183–1197

McAuliffe GA, Chapman DV, Sage CL (2016) A thematic review of life cycle assessment (LCA) applied to pig production. Environ Impact Assess Rev 56:12–22

Ministerio de Agricultura, Alimentación y Medio Ambiente (2014) Real Decreto 4/2014, de 10 de enero, por el que se aprueba la norma de calidad para la carne, el jamón, la paleta y la caña de lomo ibérico. Boletín Oficial del Estado 10:1569–1585

Nguyen TLT, Hermansen JE, Mogensen L (2011) Environmental assessment of Danish pork, p 39. https://dcapub.au.dk/djfpdf/ir_103_54761_indhold_internet.pdf

Nieto R, García-Casco J, Lara L et al (2019) Ibérico (Iberian) Pig. European Local Pig Breeds - Diversity and Performance A study of project TREASURE. IntechOpen. https://doi.org/10.5772/intechopen.83765

Pietrosemoli S, Tang C (2020) Animal welfare and production challenges associated with pasture pig systems: a review. Agric 10:223

Pineda FD, Montalvo J (1995) Dehesa systems in the western Mediterranean: biological diversity in traditional land use systems. In: Halladay P, Gilmour DA (eds) Conserving biodiversity outside protected areas. The Role of Traditional Agro-ecosystems, IUCN, Gland, pp 107–122

Plieninger T, Höchtl F, Spek T (2006) Traditional land-use and nature conservation in European rural landscapes. Environ Sci Pol 9:317–321

Plieninger T, Flinzberger L, Hetman M, Horstmannshoff I, Reinhard-Kolempas M, Topp E, Moreno G, Huntsinger L (2021) Dehesas as high nature value farming systems: a social-ecological synthesis of drivers, pressures, state, impacts, and responses. Ecol Soc 26(3):23

Reyes-Palomo C, Aguilera E, Llorente M, Díaz-Gaona C, Moreno G, Rodríguez-Estévez V (2023) Free-range acorn feeding results in negative carbon footprint of Iberian pig production in the dehesa agro-forestry system. J Clean Prod 418:138170. https://doi.org/10.1016/J.JCLEPRO.2023.138170

Rodríguez Estévez V, Mata Moreno C (2006) Alimentación de cerdos con recursos naturales y restos de cosecha. Fertilidad de la tierra: revista de agricultura ecológica 26:56–60

Rodríguez-Estévez V, García A, Perea J, Mata C, Gómez AG (2007) Producción de bellota en la dehesa: factores influyentes. Arch Zootec. 56(2):1–3

Rodríguez-Estévez V, García A, Peña F, Gómez AG (2009a) Foraging of Iberian fattening pigs grazing natural pasture in the dehesa. Livest Sci 120:135–143

Rodríguez-Estévez V, García A, Gómez AG (2009b) Characteristics of the acorns selected by free range Iberian pigs during the montanera season. Livest Sci 122(2–3):169–176

Rodríguez-Estévez V, Sánchez-Rodríguez M, García A, Gómez-Castro AG (2010) Feed conversion rate and estimated energy balance of free grazing Iberian pigs. Livest Sci 132(1–3):152–156

Rodríguez-Estévez V, Sánchez-Rodríguez M, García AR, Gómez-Castro AG (2011) Average daily weight gain of Iberian fattening pigs when grazing natural resources. Livest Sci 137(1–3):292–295

Rodriguez-Estevez V, Sanchez-Rodriguez M, Arce C et al (2012) Consumption of acorns by finishing Iberian pigs and their function in the conservation of the Dehesa agroecosystem. Agroforestry for biodiversity and ecosystem services—science and practice. InTech. https://doi.org/10.5772/34877

Rodríguez-Estévez V, Reyes Palomo C, Sanz Fernández S, Rodríguez Hernández P, López Tirado J, Sánchez Rodríguez M, Díaz Gaona C (2022a) Catálogo de Plantas Usadas Tradicionalmente para Alimentar a los Cerdos. Asociación Valor Ecológico—Ecovalia, Sevilla, p 113

Rodríguez-Estévez V, Reyes Palomo C, Sanz Fernández S, Rodríguez Hernández P, López Tirado J, Sánchez Rodríguez M, Díaz Gaona C (2022b) Materias primas tradicionales, productos alternativos y subproductos para la alimentación del porcino ecológico. Asociación Valor Ecológico—Ecovalia, Sevilla, p 125

Sánchez-Esquiliche F, Ferreira C, Rodriguez-Estevez V, Muniesa A (2023) Prediction of growth performance parameters in the growing and free-range finishing phases of the Iberian pig via meta-analysis. Span J Agric Res 21(1):e0603–e0603

Sanz-Fernández S, Díaz-Gaona C, Casas-Rosal JC, Alòs N, Tusell L, Quintanilla R, Rodríguez-Estevez V (2024) Preweaning piglet survival on commercial farms. J Anim Sci 102:1–10. https://doi.org/10.1093/JAS/SKAD408

Soussana J-F, Loiseau P, Vuichard N, Ceschia E, Balesdent J, Chevallier T, Arrouays D (2004) Carbon cycling and sequestration opportunities in temperate grasslands. Soil Use Manag 20:219–230. https://doi.org/10.1111/J.1475-2743.2004.TB00362.X

Tejeda JF, Gandemer G, Antequera T, Viau M, García C (2002) Lipid traits of muscles as related to genotype and fattening diet in Iberian pigs: total intramuscular lipids and triacylglycerols. Meat Sci 60:357–363. https://doi.org/10.1016/S0309-1740(01)00143-7

Whitehead D (2020) Management of Grazed Landscapes to increase soil carbon stocks in temperate, Dryland Grasslands. Front Sustain Food Syst 4:197. https://doi.org/10.3389/FSUFS.2020.585913/BIBTEX

Zu Ermgassen EK, Phalan B, Green RE, Balmford A (2016) Reducing the land use of EU pork production: where there's swill, there's a way. Food Policy 58:35–48

Index

A
Acorns, 168, 169, 171–178, 180–182
Agroforestry, 75
Alternative protein sources, 60–61, 64

C
Carbon footprints, 26, 32–36, 61, 75, 178, 179
Carbon sequestration, 32–34, 36, 37, 75, 178, 179
Chile, 14, 15, 128, 135, 136, 142, 143, 150–163
Chilean Patagonia, 128, 137, 139, 144
Consumer preferences, 19–26, 52, 178
Conversion challenges, 74–87, 94–117

D
Dairy production systems, 87
Dehesas, 167, 168, 171, 172, 175–179, 182

E
Ecosystem services, 75, 177, 181
Extensive grazing, 135

F
Farming, 4–6, 8, 11, 19, 31–33, 36, 37, 40, 45, 46, 51, 56, 57, 60–63, 66, 74, 78, 82–87, 98, 99, 102, 105, 109, 113, 114, 116, 117, 128, 139, 150, 153, 160, 162, 163, 168, 172, 178, 182, 183
Fatty acids (FA), 75, 76, 82, 84, 86, 87, 161, 162
Fatty pig, 63, 66, 169, 172, 174, 178
Free-range, 35, 47, 58, 64, 65, 161, 168, 174–175, 177, 178, 183

G
Grazing, 5, 32–35, 83, 86, 95, 99, 101, 102, 104–107, 109, 110, 116, 128, 131–135, 139, 141, 143, 151, 157, 158, 161, 167, 168, 171, 174–178
Grazing systems, 36, 102–106, 116, 132–135, 143, 161
Greenhouse gases (GHGs), 30–37, 75, 178, 182

I
Iberian breed, 168
Indonesia, 39–53
Insect inclusion, 61–66
Insect meals, 60–62, 66
Intention-behaviour gap, 23

L
Lamb meat, 35, 135, 143, 152, 161–163
Lamb production, 128–144
Lambs grazing, 134, 136
Life cycle assessment (LCA), 36, 182

M
Marketing of organic product, 26
Meta-analysis, 76, 78, 81, 85–87, 174, 175
Milk composition, 74–87

C. Palacios Riocerezo, M. Gonzalez Ronquillo (eds.), *Traditional Livestock Production*, https://doi.org/10.1007/978-3-031-88275-3

N
Native chickens, 45, 46, 48
Nature, 57

O
Organic, 2, 19, 34, 57, 74, 94, 128, 151, 168
Organic dairy production, 98, 104, 106, 107
Organic European regulation, 168
Organic farming, 2–9, 19, 23, 24, 33, 57–59, 66, 94, 97, 99, 102, 104, 105, 111, 113, 116, 180–182
Organic feed, 8, 9, 11, 57, 178, 183
Organic foods, 9, 10, 15, 19–26, 74–76
Organic markets, 2, 3, 10, 13, 14, 20–22
Organic products, 3, 8, 10–16, 21–25, 74, 95, 96, 163, 181

P
Production systems, 2–16, 35, 56–66, 74–76, 78, 87, 94, 96, 97, 101, 108, 109, 144, 151, 159, 162, 167–183
Purchase behaviour, 23

R
Rangelands, 128, 130–135, 137, 143, 144, 161
Reproductive efficiency, 58, 102, 108, 111–112

S
Sheep, 32–35, 77–80, 82, 84, 86, 128, 132, 133, 135–137, 139, 140, 142, 144, 150–163
Small scale business, 22, 51, 59
Sustainability, 3, 4, 6, 10, 11, 19, 22, 31, 35, 37, 57, 75, 113, 128, 150, 156, 168, 178, 181, 183

U
Udder health, 97–99, 109, 110, 116

W
Wildlife interactions, 144